CAMBRIDGE LIBRARY COLLECTION

Books of enduring scholarly value

Life Sciences

Until the nineteenth century, the various subjects now known as the life sciences were regarded either as arcane studies which had little impact on ordinary daily life, or as a genteel hobby for the leisured classes. The increasing academic rigour and systematisation brought to the study of botany, zoology and other disciplines, and their adoption in university curricula, are reflected in the books reissued in this series.

Memorials of Sir C.J.F. Bunbury

Sir Charles James Fox Bunbury (1809–86), the distinguished botanist and geologist, corresponded regularly with Lyell, Horner, Darwin and Hooker among others, and helped them in identifying botanical fossils. He was active in the scientific societies of his time, becoming a Fellow of the Royal Society in 1851. This nine-volume edition of his letters and diaries was published privately by his wife Frances Horner and her sister Katherine Lyell between 1890 and 1893. His copious journal and letters give an unparalleled view of the scientific and cultural society of Victorian England, and of the impact of Darwin's theories on his contemporaries. Volume 7 covers the years 1873–7. With advancing age there are many notices of deaths or illnesses of friends or colleagues, such as Sedgwick, Agassiz and Lyell, but Bunbury was still enthusiastically discussing his reading with correspondents.

T0188050

Cambridge University Press has long been a pioneer in the reissuing of out-of-print titles from its own backlist, producing digital reprints of books that are still sought after by scholars and students but could not be reprinted economically using traditional technology. The Cambridge Library Collection extends this activity to a wider range of books which are still of importance to researchers and professionals, either for the source material they contain, or as landmarks in the history of their academic discipline.

Drawing from the world-renowned collections in the Cambridge University Library, and guided by the advice of experts in each subject area, Cambridge University Press is using state-of-the-art scanning machines in its own Printing House to capture the content of each book selected for inclusion. The files are processed to give a consistently clear, crisp image, and the books finished to the high quality standard for which the Press is recognised around the world. The latest print-on-demand technology ensures that the books will remain available indefinitely, and that orders for single or multiple copies can quickly be supplied.

The Cambridge Library Collection will bring back to life books of enduring scholarly value (including out-of-copyright works originally issued by other publishers) across a wide range of disciplines in the humanities and social sciences and in science and technology.

Memorials of Sir C.J.F. Bunbury

VOLUME 7: LATER LIFE PART 3

EDITED BY
FRANCES HORNER BUNBURY
AND KATHARINE HORNER LYELL

CAMBRIDGE
UNIVERSITY PRESS

CAMBRIDGE UNIVERSITY PRESS

Cambridge, New York, Melbourne, Madrid, Cape Town,
Singapore, São Paolo, Delhi, Tokyo, Mexico City

Published in the United States of America by Cambridge University Press, New York

www.cambridge.org
Information on this title: www.cambridge.org/9781108041188

© in this compilation Cambridge University Press 2011

This edition first published 1892
This digitally printed version 2011

ISBN 978-1-108-04118-8 Paperback

MEMORIALS

OF

Sir C. J. F. Bunbury, Bart.

EDITED BY HIS WIFE.

THE SCIENTIFIC PARTS OF THE WORK REVISED BY
HER SISTER, MRS. LYELL.

LATER LIFE.

VOL. III.

MILDENHALL:
PRINTED BY S. R. SIMPSON, MILL STREET.
MDCCCXCII.

1873.

JOURNAL.

January 1st.

A beautiful bright day, as mild and enjoyable as 1873. spring. Spent much of the morning out-of-doors with Fanny, Cecil and Herbert and the dogs.

January 3rd.

The children's dance and merry-making—a large party and much merriment.

January 6th.

Resumed my father's Hawkwood MSS.

Tuesday, January 14th.

Minnie Powys arrived—also Lord Waldegrave and Mr. St. John. Minnie Powys sang to the guitar.

LETTER.

Barton, Bury St. Edmund's,
January 14th, '73.

My Dear Mary,

Last week I spent a good deal of my time

A 2

1873. in reading "The Memorials of a Quiet Life." The
greater part of the first volume pleased me very
much : I thought it both interesting and amusing,
especially the account of Hurstmonceaux and the
Hares, and all relating to Reginald Heber and
Bishop Stanley's family. In the latter part there
was too much sermonizing for me, and I skipped a
good deal, but still there is much that is interesting
and touching in the picture of the quiet happiness
of their short married life. The second volume I
cannot get on with.

Your letter written on Christmas Day gave me
great pleasure. You were then reading "Middle-
march:" I like it better on the whole than "Adam
Bede," though there is nothing in it so good as Mrs.
Poyzer. The heroine I think charming, a beautiful
character, and the contrast between the two sisters
is very good and very well kept up, but I cannot
at all reconcile myself to Dorothea's marrying
Ladislaw, it makes me quite angry that she should
take that young scamp for her husband : and it is
so evident that she marries him just because her
first husband had tried to prevent her doing so.

(January 21st). I am still very weak and my
cough not quite gone, I am much better. I was
particularly unlucky in being unwell last week, for
we had a very pleasant and gay party in the house,
chiefly young people, and I was not able half to
enjoy the society: however, they all seemed to be
very merry together, and Clement supplied my
place very well.

So the Emperor Napoleon is dead : there is an 1873.
extraordinary man gone :—he who for so many
years filled so great a place in the thoughts and
calculations of all public men throughout Europe,—
he to whom all looked as to a destiny :—where are
his schemes and projects now? " Poca polvere son,
che nulla sente." How many thousand lives would
have been saved if he had died three years ago ? I
am glad, though, that he ended his days peacefully
(as far as his malady would allow) among us.
The French must feel that Chiselhurst makes
some amends for St. Helena.—And now Lord
Lytton is gone, who I think will always be reckoned
among the great novelists of England, and Dr.
Lushington too, who I should think must long have
out-lived all his old colleagues and opponents.

With much love to Charles Lyell,

> Believe me ever,
>> Your affectionate brother,
>>> CHARLES J. F. BUNBURY.

JOURNAL.

January 15th.

Fanny took me out in the brougham in the
morning, we drove round by Rougham.

The Robert Anstruthers, the Montgomeries, Mrs.
Byng and her two daughters and several young men
arrived.

1873.
January 16th.

Had another little drive in the morning with Fanny. The whole party except me went to the Bury ball.

January 17th.

A beautiful day. I had a little walk in the pleasure grounds. The Wilsons and Hortons came with numerous parties—our dance very gay and well kept up—a very pretty sight.

January 18th.

All our guests went away, including Clement. I had a little stroll in pleasure ground.

January 20th.

A fine cold day, white frost in morning. Took a drive with Fanny, round by Ingham and Livermere. Wrote to Sarah Hervey.

January 23rd.

On the 7th I was attacked with a feverish cold, which though it did not at first threaten to be very severe, proved very obstinate and troublesome, has worried me ever since, has weakened me very much, and for a good while made me incapable of writing, or of using my mind much. Nor am I yet quite well.

During this interval an extraordinary man has departed this life; the Emperor Napoleon the Third is dead. I have not time to write down half the thoughts which such a death suggests. It remains for posterity, for history to judge him. He certainly was not as great, as supreme in any one department, as his uncle was in the art of war; yet he made himself for several years, almost as much the arbiter of Europe, almost as much the centre of all attention and all political speculation, as ever his uncle was. Whatever he was as towards France, we English had every reason to speak well of him in his relation to us; he was always our good friend and ally; and the feeling shown on the occasion of his funeral at Chiselhurst, the other day, evinced a proper sense of this. If his fall was almost as great as that of his uncle, there was more to soothe his last hours. Rather a contrast between Chiselhurst and Longwood.

Since then Lord Lytton has gone, who I think will hold a permanently high place among English writers of fiction; and Dr. Lushington, one of the last remaining links between these days and the early part of the century.

_____

January 24th.

From the 14th to the 18th we had a pleasant party in the house, but I was too unwell to be able half to enjoy the company. The ladies were all agreeable, and most of them pretty: Mrs. Anstruther and Mrs. Montgomerie remarkably pretty.

_____

January 25th.

Dear little Arthur MacMurdo left us to go to

1873. school at Bengeo, near Hertford, a preparatory school for Haileybury. For nearly two years he has been almost like an adopted child to Fanny, who has taken the whole charge of his education : we shall miss him much.

I walked about the arboretum where I had not been since the 5th.

January 27th.

Mr. and Mrs. Drummond, the Barnardistons, Mrs. and Miss Eyre arrived. Lady Cullum, the Wilsons and the Hortons dined with us.

January 28th.

Yesterday we received the news of Adam Sedgwick's death. His niece wrote to Fanny, that it was a most peaceful and happy death-bed : that he suffered no pain, only weakness, slept much, and lay in general perfectly quiet. Sedgwick, it appears, had nearly completed his 88th year, having been born in 1785. Only a very short time ago, he sent my wife a present of his last publication, a Preface to the Catalogue of Cambrian Fossils in the Woodwardian Museum, with a touching address to her in his own handwriting. Yesterday afternoon a postal card from his faithful servant informed us that he had passed away very peacefully early that morning. So there is another noble veteran of science gone from among us. He was not only a geologist of a very high order, but a man of very remarkable general force of intellect and character,

and moreover, an excellently good man. He was 1873.
a very old friend to both of us, especially to
Fanny, who had known him from her childhood,
and he was repeatedly our guest, a very agreeable
one at Mildenhall. When he was in good health
and in congenial company, his talk was extremely
entertaining and instructive. There was a native
vigour and robustness in his intellect and character,
a sort of rocky sturdiness (corresponding to the
granite-like air of his head), a hearty geniality and
strength, combined with the greatest tenderness of
heart, which was extremely interesting and delightful. His memory was rich in old stories of the
people of his native Yorkshire dales, which he told
admirably well. He had great powers both of
humour and pathos. Sedgwick was not only
eminent as a man of science, but had great power
as a writer of English, and might have made for
himself a great name in literature. I remember
my uncle, Sir William Napier, pointing out to me
Sedgwick's first Presidential Address to the
Geological Society, and expatiating on the extraordinary merit of the style. Sir William also, I
remember, once said that Sedgwick, if he had
chosen the army as his profession, would probably
have been a great commander. In fact he had
so powerful and vigorous an intellect, as would
probably have enabled him to excel in almost
anything that he had taken up in earnest.

_____

February 3rd.

The newspapers contain a very interesting

1873. account of the funeral of dear old Sedgwick in the Chapel of Trinity College, where I have so often seen him. It must have been a fine and touching scene.

Dear Joanna Horner arrived, looking extremely well: she has not been in England since the autumn of 1870.

February 4th.

My 64th birthday—thanks be to God. Received very kind letters from Cissy, Mary, Katharine, Leonora and Edward.

======

LETTERS.

Barton, Bury St. Edmund's,
February 4th, 73.

My Dearest Cissy,

I thank you most heartily and sincerely for your kindness in sending me those delightful drawings of my father's, and accompanying them with such a very kind and loving letter. No gift could have been more welcome to me than those drawings; they are capital in themselves, and several of them are especially precious to me, because I remember so well seeing my father drawing them, and they are so much associated in my memory with former days. I prize them exceedingly.

Fanny has had a fine book made, in which the

drawings you had already given me are mounted, 1873. and these will be a precious addition to the collection. But the expressions, so warm and cordial of your love and kindness to me are more delightful than the drawings, and have gone to my very heart, dear Cissy; and be assured I most heartily reciprocate all your and Henry's kind feelings towards me. I have a great deal to be thankful for, and scarcely anything more than having such kind and true and loving friends. It distresses me exceedingly to hear that dear Henry is suffering so much pain which I much fear must prey upon his health and strength.

I am very glad that we shall be in town at the same time with you, and hope we shall see much of you.

With my best love to dear Henry and Emmy,

Believe me ever,

Your very affectionate brother,

CHARLES J. F. BUNBURY.

Barton, Bury St. Edmund's,
February 6th, 73.

My Dear Edward.

I thank you very much for your kindness in writing to me, and for your good wishes on my birthday. I fully go along with you in what you say about the conditions which can alone make a great age desirable, and about those examples of happy long life which have so lately been brought to our thoughts. Though Sedgwick at times suffered

1873. a great deal from ill health, he preserved his faculties, I believe,—certainly those of his mind to the last, and with such a union of mental activity and vigour with extreme kindness and warmth of heart, he must have been happy. His funeral in Trinity Chapel must have been an interesting sight. It is certainly remarkable, as you say, how many remarkable persons have lately passed away.

Joanna Horner is staying with us, which is a great pleasure.

Ever your affectionate brother,
Charles J. F. Bunbury.

Barton,
February 7th, 1873.

My Dear Leonora,

Very many thanks for your kind letter and good wishes on my birthday. I may well be thankful for having such comparative good health and so much power of enjoying life at the age of 64, and above all for such kind and delightful friends. I suffered during last month from a cold and cough, which hung about me very pertinaciously, and pulled me down a good deal, but I am now pretty well. The weather however has become detestable; after continuing so marvellously mild to near the end of January, it has turned bitterly cold, with snow, east wind, and all manner of abominations; but I do not think there has yet been enough severity of frost to allow of skating.

Fanny had an unlucky fall the other day, while adventurously climbing to catch one of her little birds which she had let out of its cage ; she bruised 1873. herself a good deal, but she seems to be getting decidedly better, and I am in hopes from what Image tells me, that she will be well, in a few days. It is a great pleasure to have dear Joanna here with us, and she is looking remarkably well and seems in excellent spirits. I am reading her and Susan's "Walks in Florence," which as far as I have gone, appears extremely well done, and is very interesting.

(February 10th). Fanny was very much disturbed this morning by the news that her pet Arthur MacMurdo, who went to school only a week ago, has been attacked with measles: but I hope he will soon get over it, as a later telegram gives a very favourable account of him, and if he has it mildly it will be rather a good thing to have got it over early in life.

I am flattered by hearing that you are reading my Cape book : it is a slight record of a time which I passed very agreeably, but I often wish I had to write it over again, as I think I could do it better. I am still reading Grote's "Greece," a long job, but it is worth the time it takes, worth reading carefully, it is a masterly work, and the history it relates is that of a most interesting and wonderful nation. I have just finished the tenth volume, and there are two more yet. I am also reading Froude's "History of the English in Ireland in the Eighteenth Century,"—very interesting, admirably well written, but a most melancholy history of nearly every blunder, folly and crime that a Government on one

1873. side and factions on the other can contrive to commit. One cannot for a moment suppose Mr. Froude to be an impartial writer, for his dislike to the Irish and his hatred of Popery are so vehemently shown as to be sometimes almost shocking, yet he *shows up* all sides with great severity.

Dear old Sedgwick's death was a fitting close to his noble life, and the description of his funeral was very interesting. There was a most excellent notice of him by Professor Phillips in *Nature* of last week. How fast the great men of the last generation are dropping off! He was one of the last of those who belonged to the " Heroic Age" of Geology.

Ever your affectionate brother,
Charles J. F. Bunbury.

JOURNAL.

February 8th.

My Barton Rent Audit — very satisfactory: luncheon afterwards to my tenants and Mr. Percy Smith.

Monday, February 10th.

Fanny and Joanna went over to Mildenhall and returned to dinner.

February 13th.

Walked about the arboretum. Arranged some

books. Read the last four chapters of the Book of 1873.
Judges, with Lord Arthur Hervey's Commentary.
News of the abdication of the King of Spain. I
think he is quite in the right.

February 14th.

My nephew Cecil was married on the 11th to
Susan Napier, William's second daughter. God
grant they may long be happy together. She is a
charming and admirable girl, and he is a fine, noble
minded, excellent young man: morally, nothing
could be desired better for a happy union: but
what makes one apprehensive is, the frightfully
delicate state of her health.

February 15th.

Up to London by S. Pancras, and then to 48,
Eaton Place, Joanna with us, arrived safe and well,
thank God. A visit from Edward. A very fine and
pleasant day, which has been a very great rarity
this month.

February 16th.

Henry, Cissy and Emily, Charles and Mary
Lyell, Sarah Seymour, Sir Edward Ryan and John
Carrick Moore came to see us. Henry seems
rather better in health than I had apprehended;
Charles Lyell, I am sorry to say, looks very ill.

Sir Edward Ryan approves much of the
Chancellor's Grand Bill for a Supreme Court of
Judicature.

February 18th.

Joanna Horner left us to go and stay with Katharine. She has been a most agreeable companion, and is indeed an admirable woman. I received a very interesting letter from Sarah Hervey. She is going abroad immediately with her sister and brother-in-law, to the Pyrenees. (This plan has since been altered, wisely, and they are going to Nice and Riviera, a much more suitable expedition for this season).

February 19th.

Fanny went off to Bengeo School, near Hertford, and saw Arthur MacMurdo. She returned at 6. Clement dined with us. Dear Kate and Charley Hoare came in to see us for a few minutes in the evening.

Friday, 21st.

Lady Bell came to luncheon. Called on the Lady Legges, found Wilhelmina alone at home. Yesterday I had the great pleasure of seeing Charles Kingsley and his two daughters Rose and Mary. They came to luncheon with us. This was a candle-light luncheon party, the fog being so dense and dark at 2 o'clock, that we thought it best to light the candles and shut the shutters of the dining room as if it were night. Mrs. Kingsley unfortunately was unable to come, being ill with a feverish cold. Kingsley is looking very well, and appears quite unchanged since he was last at Barton, 4 years ago:

he is as agreeable as ever, and appears as energetic
and active.

Kingsley tells me that his son Maurice has been
engaged in exploring and surveying for a new rail-
road, which is designed to run through the northern
provinces of Mexico, from the city of Mexico, to
join the great American railway which runs across
the state of Colorado and the Rocky Mountains
to the Pacific. A gigantic scheme.

I was particularly glad to meet his daughter Rose,
whom I have not seen once in these last four years,
and who has accomplished such an adventurous
journey across North America to the state of
Colorado, and again across Mexico from sea to sea.
She is noble, both in intellect and character, and
I was delighted to find that her American experiences and her rough travels had not in the least
spoiled her agreeable qualities ; had given no tinge
of coarseness or of American peculiarity to her
manners, nor induced the least appearance of anything hard or unfeminine. In short, she is as
agreeable as ever, and more than ever interesting.
She is singularly like her father in face, and in mind
also I think. She said the people of Mexico are
remarkable for their skill and ingenuity in decorative
art, they still make up most exquisite pictures of
feathers, just as their ancestors did in the time of
Cortez. The real Aztecs have features much
resembling those exaggerated profiles which one
sees in the ancient Mexican pictures and sculptures,
but of these people there are but few remaining,
she saw them only near the lake of Tezcuco. The

1873. country is so insecure, that, except in the cities, no one can go out alone or unarmed. She spoke with enthusiasm of the beauty of the scenery and vegetation in the island of Trinidad, which she had visited with her father in 1870 : she saw no forest scenery in Mexico comparable to it. The great table-land of Mexico is (as I was aware) very bare, and on the road by which she travelled by the West Coast, the timber had been nearly all cut down. The woods about Jalapa, on the descent from Mexico to Vera Cruz, of which Humboldt speaks so much, are now only *second growth* woods, the successors of the primeval forests which have been felled ; in one part of that tract, however, the climbers are very fine. On the table-land there is no conspicuous vegetation except Cactuses (very large and strange) and Yuccas.

The cold of winter in the elevated valley or plateau where Rose was living with her brother, in the state of Colorado, is at times intense : she had known it as much as 170 below zero, and she was living in a wooden house of which the timbers were far from fitting closely. But the extreme dryness of the air renders the cold endurable. She remarked as a curious thing, that sometimes, when a considerable quantity of snow had fallen, it would all disappear by evaporation, in a short time, without melting.

Rose Kingsley says, that *sham* Mexican antiquities are manufactured in a large scale in the city of Mexico, and with great skill, and are sold at great prices to foreigners. It is quite a branch of industry.

Yuccas are a characteristic form of vegetation, 1873. both in Colorado and on the table-land of Mexico within the tropics: but in the former country they only grew three or four feet high, in the latter twenty or thirty.

Saturday, February 22nd.

Henry, Cissy and Emily came to luncheon. Dear Sarah Hervey arrived. The Albert Seymours and George Napier dined with us. Octavia and Wilhelmina Legge came in the evening.

Monday, February 24th.

Dear Sarah Hervey came to us late in the afternoon of the day before yesterday (Saturday)—spent the greater part of Sunday with us, and after afternoon Church we went with her to 37, Fleet Street, saw the Charles Hoares, and took leave of all the three, as they were about to start for Dover on their way to Nice.

Lady Louisa Legge, Sir W. Boxall, Katharine and Joanna lunched with us on Sunday. Henry lunched with us. We dined with the William Napiers in Pembridge Square.—Strictly a family party.—Cecil and Susy.

February 25th, Tuesday.

Thaw and rain. Minnie came to luncheon, and Cissy and Emmy afterwards, then Charles and Mary.

B 2

1873. We went yesterday to the Exhibition of the
Works of the Old Masters—a very interesting and
beautiful collection. I was particularly struck, in
this first visit, with several fine landscapes by
Crome, especially, View on the Edge of a Park,
(belonging to Fuller Maitland): the Yarmouth
Water Frolic (lent by Canon Selwyn): Bruges on
the Ostend River (No. 96): Slate Quarries (lent by
Mr. Maitland). Also the portrait of Lady Anna Len-
nox, afterwards Countess of Albemarle, by *Reynolds*:
Mrs. Stanley, by *Reynolds*: a Lady with a Parrot, by
Rembrandt: a Knight putting on his Armour, by
Paris Bordone: the portraits by Romney of Mrs.
Drummond Smith (a lady in white, full face, with
an enormous hat, a most exquisite picture), and
Mrs. Carmichael Smyth, most lovely: Sir Henry
Raeburn's portrait of Henry Erskine—admirable:
a Lady opening a Casement, by *Rembrandt*: a
Shipbuilder and his Wife, by *Rembrandt*: Daniel in
the Lion's Den, by *Rubens* (the lions magnificent,
the prophet absolutely disgusting): Erasmus and
his friend Ægidius, by *Holbein* (from Lord Radnor's
collection): the Cornaro family, by *Titian* (this I
remember seeing some years ago in the Brit. Inst.):
John Hunter, by *Reynolds* (the portrait so well
known from engravings).

February 26th, Ash Wednesday.

Much milder—some rain.

We dined yesterday at Sir Edward Ryan's. Sir
Edward himself is a very fine old man in his 80th

year, of extreme knowledge and eminent ability, 1873.
very cheerful, amiable and pleasant. The Louis
Mallets, both of them very agreeable. Minnie
dined with me. Fanny went to a great concert at
the Albert Hall, and came home very late.

February 27th, Thursday.

Emily Napier, Cecil and Susie came to luncheon,
afterwards we went out with them, and spent a long
time looking at jewelry at Collingwood's and
Lambert's.

Friday, February 28th.

Went to see Henry, who seemed rather better
than I expected. Minnie, Mr. Walrond, John
Herbert, Edward and Clement dined with us.

Saturday, 1st March.

A very pleasant visit from Mr. Gambier Parry.

Monday, 3rd March.

A very pleasant visit from my cousin Norah
Bruce : she is a very agreeable and most excellent
person, very like her admirable mother, my aunt
Caroline. She seems to be pretty well recovered
from her long illness, looks well, and is grown fatter.

Went to a party at Katharine's : met there Mary,
Joanna, Cissy and Emmy, Mr. and Mrs. Arthur
Milman, &c.

Sally and her daughter Fanny, Mr. Jeune,
Edward and Clement dined with us.

Tuesday, 4th March.

The opposition to the Ministerial measure on the
Irish Universities seems very formidable, and I
should think the Government very far from secure
of victory. Mr. Fawcett's speech against the Bill,
on this first night of the debate, was very powerful
and effective indeed.

Minnie and the Albert Seymours, Cecil and
Susie, Emmy and John Herbert came to luncheon.

_____

Wednesday, 5th Mach.

Went to see Henry, and found him a little better.
Then went to see Charles and Mary Lyell, and had
a very pleasant talk with them.   Charles seemed
unusually well, and talked with more animation and
with less appearance of effort than I have heard him
for some time.   He had been much struck by the
Duke of Argyll's Presidential Address to the
Geological Society—delighted with his eloquence,
and impressed by some of his reasoning.

Edward dined with us: he says that he finds that
all the Members of Parliament with whom he has
talked, think that the Ministry cannot carry their
Irish University Bill in its present form.

_____

Friday, 7th March.

Henry certainly better.   Minnie Powys came to
luncheon.   Charles Napier and Minnie dined
with us.

_____

Saturday, 8th March.

Fanny went to Hertford to see Arthur, and
returned to dinner.

Minnie Napier, Sarah and Albert Seymour, 1873. William Napier, Edward and Cecil dined with us.

Monday, 10th March.

Farewell visits from Minnie, Sarah, Henry and Cissy and the Charles Lyells.* We dined with the William Napiers. Edward the only guest.

Tuesday, 11th March,

Down to Barton by the 11.30 train from St. Pancras.—A comfortable journey—arrived safe and well, thank God, at our dear home. Fanny heard yesterday morning from Lady Arthur Hervey, that her third daughter, Patience, is engaged to be married to a Derbyshire gentlemen, Mr. Morewood. Lady Arthur seems very much pleased, and I rejoice at any thing which contributes to the happiness of such very dear friends. This Mr. Morewood (whose father is very lately dead) is the great, great nephew of an old Mr. Morewood, in whose house I (a child of 11, travelling with my father and mother) spent some time in the spring of 1820. The old gentleman and lady were excessively good-natured and kind to me, and I have a lively remembrance of the riches of their library and garden.

---

* The last time we saw my dear sister, Mary Lyell, before I saw her unconscious on her death-bed.—(F. J. B.)

March 13th, Thursday.

Mr. Allington, the School Inspector arrived. Mr.
Percy Smith dined with us.

———— ——

Friday, March 14th.

News of the resignation of the Gladstone
Ministry.

———— ——

Saturday, March 15th.

It is said that D'Israeli has declined to undertake
the Government. He is wise, I think.

Examined the structure of the flowers of Phaius
grandifolius, now in flower in our conservatory, and
made a note of it. This is a stately and beautiful
Orchid.

———— ——

Tuesday, March 18th.

Gladstone is engaged in re-constructing his
Ministry.

———— ——

Wednesday, 19th March.

Fanny went over to Mildenhall and returned to
dinner.

Examined the structure of the flowers of a very
beautiful Dendrobium nobile (?) which is in flower
in our conservatory, and made a note of it.

———— ——

Saturday, 22nd March.

A pleasant visit from Lady Bristol and Lord
Henry Percy.

Received an interesting letter from Sarah Hervey. 1873.
They have returned to Wells, having been obliged
to stop short at Paris, and give up their southern
tour on account of Kate's health. But Sarah
visited with her brother-in-law the principal places
in the neighbourhood of Paris, and sends me
interesting remarks on Versailles, St. Denis, St.
Germain and Fontainbleau.

LETTER.

Barton, March 26, '73.

My dear Susan,

I hope you have now the pleasure of having
Joanna at home with you again, and that she is not
the worse for her winter journeys; it is a very good
thing they are over; I wondered at her courage
in undertaking them at such a season. Certainly
she was looking remarkably well when she was with
us, and was exceedingly pleasant; I was very sorry
to part with her.

I have read the greatest part of your joint book,*
not all straight through; though it bears reading
even in that way much better than one could have
expected a work of that kind to do; but the
pleasantest way, I think, is to take up one after
another of the most interesting localities, and read
what belongs to each as a separate chapter. I like
the book extremely; the amount of research and
industry it shows is is quite surprising; and the
historical illustrations and anecdotes are very skil-

* Walks in Florence.

1873. fully and agreeably introduced and very well told ;
they relieve very pleasantly the mass of local descriptive details. That association of striking local
features with memorable historical incidents mu-
tually throwing light upon one another, is indeed
remarkably characteristic of Florence, and is extremely well brought out in your book. I hope
it will have the success which it well deserves. I
am still engaged in reading Grote, and am now
in the middle of the 11th volume, and in the time of
Philip of Macedon—just come to Demosthenes and
Phocion. It is certainly a very long book, and
perhaps some parts here and there might be
shortened with advantage ; but, long as it is, it still
is well worth reading, and reading with care ; and it
is a book that one ought to be the better for.
Epaminondas and Timoleon—and perhaps I ought
to add Pericles, appear to me the greatest and
noblest of all the Greeks. But altogether they were
a wonderful and a glorious people, and it will be
a bad sign for mankind if their history is ever
forgotten or neglected. I have lately read Froude's
" History of the English in Ireland in the eighteenth
century"—the one volume which is as yet published ; a very strange and startling book. Mr.
Froude has very great power as a writer, and
especially an extraordinary genius for narration—for
telling a story—equal to Walter Scott, and there
are in this book some splendid specimens of that
talent. But his dislike to the Irish and to the
Church of Rome is carried to an excess that really
sometimes seems hardly sane. His dislike to the

common and generally received ways of thinking in 1873. the present day carries him so far, that he actually seems to recommend religious persecution, and to maintain the right of a powerful nation to subdue its weaker neighbour.

I read Milman's "History of the Jews" a few years ago; I am much inclined to agree with you as to the national character of the Jews; but in their adversity they were very grand; and I think (though it is not the orthodox opinion) that they were very right in resisting the Romans to the utmost of their power.

I take it for granted that Fanny has written you all our news,—the marriages, &c. We have just seen in the newspaper that Mr. Kingsley is to be a Canon of Westminster, which I am very glad of, as it will bring him and his family to London for part of the year, instead of sending them out of the way to Chester. Fanny will have told you of our seeing him and his daughters in town. Rose is as agreeable as ever; quite unspoiled, not grown hard or masculine from her adventurous travels to the far west of America with her father, which I had rather feared might be the case. She is beginning to relate her travels very amusingly in *Good Words*. We have been enjoying most beautiful weather for several days past, and it has quite removed the disagreeable impression of the long winter. I hope you and Joanna are enjoying equally agreeable weather in beautiful Florence.

Ever your very affectionate brother,
CHARLES J. F. BUNBURY.

JOURNAL.

March 28th, Friday.

1873. My niece Fanny was married yesterday to Mr. Frederick Jeune, younger son of the late Bishop of Peterborough. They are a handsome couple.

We drove to Hardwick, saw Lady Cullum, and walked about with her for some time, enjoying the fine weather and the beauty of those grounds. She has certainly shown extraordinary skill and taste in laying them out. It is wonderful how much she has made of such a tame country, by judicious planting, and opening of views, by well contrived avenues and skilful peeps of distance. She has a real talent for landscape gardening.

We have lately had information from Katharine that her son Leonard is engaged to be married to a Miss Mary Stirling. The Abrahams, Lady Cullum, Captain Horton, Mr. and Mrs. Bain, Francis Anstruther, and Mr. Allington dined with us.

March 30th.

Mrs. Abraham has sent me a copy of an epigram on Mr. Lowe*, which has been sent to her from London :—

EPITAPH ON THE RT. HON. R. LOWE.

This is the tomb of Robert Lowe,
Where he is gone no man may know,
If he has soared to realms above,
There is an end of peace and love ;
If he has found a lower level,
Then we must pity the poor devil.

* We knew Mr. Lowe afterwards, and thought him very agreeable, and of very good opinions.—F. J. B.

Tuesday, 1st April. 1873.

A very fine day.

We took a long drive in the open carriage by Pakenham, Norton, Tostock, Beyton and Thurston.

Wednesday, 2nd April.

Another fine day. Lady Young and Alice Praed came to luncheon.

Received a very pleasant little letter from Charles Kingsley, to whom (or rather to Mrs. Kingsley) I had written to congratulate him on his appointment to be a Canon of Westminster. He writes in a very happy tone, much pleased with the position.— He writes as follows :—" Instead of looking on it as " an earnest of future preferment, I acquiesce in it " as all I want.—What better fate than to spend " one's old age under the shadow of that glorious " Abbey, and close to the highest mental activities " of England, with leisure to cultivate myself, and to " write if I will, but deliberately, not for daily bread. " It cannot be better than it is."

Friday, 4th April.

Attended and took a part in a Meeting of Board of Guardians at Bury, to discuss the amalgamation of the Bury with the Thetford Union.

Saturday, 5th April.

We arranged books in Porch Room. Kingsley gazetted as a Canon of Westminster, and Shafto Adair as Lord Waveney of South Elmham.

LETTER.

1873.

My Dear Katharine,

Thanks for your letter. I am very glad to hear of Leonard's marriage engagement, because it gives you so much pleasure, and I rejoice at anything that contributes to your happiness. I am very sure that there is nothing which contributes so much to a man's permanent welfare, nothing which is such a help to goodness and happiness, as a wise and well-assorted marriage: and I have no doubt that Leonard has all the qualities necessary to happiness in that condition. I am quite unacquainted with Miss Stirling, but as you and Mary both think so highly of her, I am sure she must have great merits, and I trust that she is worthy of her intended husband. I believe you have always been an advocate of early marriage, for a man: it is a point on which there is much difference of opinion, but I hope in this instance at least, it will turn out quite as you could wish:—indeed, if I rightly understand, it is not intended that the actual marriage should take place very soon. I hope Charles and Mary and their companions are enjoying their stay at Ludlow, and that they have fair weather for their excursions. Ludlow must be an interesting place in many ways, and in a beautiful country I should think: I wish I had seen it.

Here, after a week of beautiful and enjoyable

weather, it has turned very cold and stormy again, 1873.
and this morning it is hailing or sleeting as if a new
edition of winter were beginning.

Our "spring garden" is in great beauty already
with hyacinths of many colours, primroses, forget-
me-nots and arabis, but the tulips are not yet out.
The winter, though tedious and disagreeable, has
not been severe enough to hurt vegetation, and all
our out-door trees and shrubs are thriving.

Dr. Tyndall's lecture on Niagara must have been
interesting.

I have been looking into Dr. Wyvill Thomson's
"Depths of the Sea," in which there are a great
number of most beautiful illustrations, representing
the wonderful deep-sea creatures, and most strange
fantastic looking monsters many of them are. I
find a great deal of entertainment in reading
Hayward's Essays, which are Republications of
Review articles.

What a terrible catastrophe that is of the
Atlantic,—and very disgraceful it would appear
according to our present knowledge.

Believe me ever,

Your affectionate brother,

CHARLES J. F. BUNBURY.

JOURNAL.

Tuesday, 8th April.

Dear little Arthur arrived from school. Mrs.
Wilson and her daughter came to luncheon with us.

Wednesday, 9th April.

Sent to Professor Humphrey my subscription to the Sedgwick Memorial Fund.

Thursday, 10th April.

George Bunbury arrived in the morning and Willie in the afternoon.

April 11th, Friday.

The 29th anniversary of our happy betrothal.

Monday, 14th April.

A beautiful, bright, and really warm day—quite delicious. Strolled with Fanny.

Mr. Wilson and Lady Hoste came with a party of children, and the boys all played cricket.

Wednesday, 16th April.

The Times of yesterday announced the death of my cousin General Fox. I had heard some months ago that he had fallen into a very bad state of health. He was a good and very amiable and a very pleasant man. I have not seen him very often of late years (though we have called on him every season when we were in London), but whenever we have seen him, he has always greeted us with the most friendly cordiality. He had in an eminent degree, that charm of cordiality, both in manner and character, that frank, hearty, genial kindness,

for which his father has been celebrated by 1873.
Macaulay. He was, I believe, a man of considerable knowledge and accomplishment: a good
scholar, and especially learned in Greek coins.

I wrote to Lady Lilford. Another beautiful day.
We drove to the shrub, and admired the anemones
and primroses.

————

Friday, 18th April.

Captain and Mrs. Horton and the two Miss
Brokes and Lady Cullum dined with us.

====

LETTER.

Barton,
April 22nd, '73.

My Dear Edward,

You will be sorry to hear that poor Henry
Blake (not Sir Henry, but his son) died yesterday,
quite suddenly, from the rupture of a blood-vessel
on the brain. His death was caused (Image tells
me) by the results of excessive fasting during Lent,
together with excessive clerical work, for all which
his constitution was much too weak. We have not
seen very much of him since we have been living
here, but he was very friendly when we did meet.
He was a very good man. It is sad that poor old
Sir Henry, who has been such a sufferer for so many
years, should have out-lived his son.

I was very sorry for the death of General Fox,
though he was not quite unprepared. You must

c

1873. have felt it much, as you have seen so much more of him of late years. He was always exceedingly friendly and pleasant when we did meet, and there was something very delightful in his frank, cordial, genial manners.

We are both of us pretty well, but in a good deal of anxiety about Mary Lyell, who is ill with a low gastric fever, a very disagreeable ailment. Happily it is, we are assured, of a mild kind, all the symptoms are favourable, and she is going on as well as can be in such a malady, but I am afraid it will be a good while before she is well again. This illness drove them back from their tour, and as there are workmen in their house in Harley Street, they are staying in the Great Western Hotel, but she is not able to see any one.

George and Willie are staying with us just now, as well as little Arthur. We propose to go up to town about the 15th of May, and I hope we shall find you well.

With Fanny's love,

Believe me ever,

Your affectionate brother,

CHARLES J. F. BUNBURY.

JOURNAL.

Wednesday, 23rd April.

Up to London suddenly on account of an alarming telegraph of Mary Lyell's dangerous illness.

LETTER.

48, Eaton Place,
April 25th, 1873.

My Dearest Cissy,

You will have heard already (for I think 1873. they telegraphed to you from Barton) of the blow that has fallen upon us, the sad, sad termination of dear Mary Lyell's illness. We were not without hope even so late as yesterday morning; Fanny who had spent all the night with her, came back here to breakfast with a slight better account, but even while she was at breakfast, came a message that Mary was worse. Then I felt there could hardly be any more hope; Fanny hurried back, but all was over before she arrived. Mary passed away very tranquilly—almost insensibly: and, indeed, she seems (from what I hear) to have suffered hardly anything except weakness, the whole time. I need not tell you, who knew her, what a noble creature she was. It is, indeed, a loss never to be forgotten, or made up. I loved her very dearly and with good reason, and had a great admiration for her fine talents and still finer character. To poor Charles Lyell it is indeed a crushing blow, an irreparable incalculable loss: she was everything to him,—of late years above all, since he has been broken in health and almost blind. It seems so strange—almost inconceivable, and I had never contemplated the *possibility*—I might almost say—of *her* passing away before *him*. It is a most mysterious dis-

G 2

1873. pensation, but we can only submit, in the faith that all is ultimately right, though it is hard to bear.

Katharine and her children will be immense comforts to Charles Lyell: I hope *her* health will not break down, but she looks wretchedly ill.

Fanny is as well as she can be under such a blow, and I am happy to say she had a good sleep last night.

Much love to Henry and Emmie, and believe me,

Your very affectionate brother,

CHARLES J. F. BUNBURY.

JOURNAL.

May 30th.

The 29th anniversary of our happy marriage.— God be thanked.

May 31st.

Since the last entry in this journal, we have passed a melancholy time. We saw dear Mary Lyell before we left London in the beginning of March looking well, in excellent spirits, and apparently free from any malady. She and Charles had arranged a plan for a geological expedition to Ludlow, in company with their nephews Leonard and Arthur, and she seemed to be looking forward to it with pleasure. Towards the end of March we heard that she had an attack of influenza—such it was then supposed to be, but there was nothing in the accounts to alarm us, and, indeed, neither she herself nor those about her apprehended that it was 1873. anything more than an ordinary attack. They went to Ludlow, as had been planned, about the 2nd of April, and stayed there till the 14th, but during this time her languor and debility (which had at first been attributed to the influenza) continued and increased, and at last excited uneasiness. The party returned to London on the 15th, but as there were workmen in the house in Harley Street, she and Charles stopped at the Great Western Hotel. She at once took to her bed, from which she never again rose. Still, though her malady was now acknowledged to be a low gastric fever, still, neither she herself nor those about her, would believe that she was in danger; we were constantly assured that the symptoms were satisfactory, and that she was going on perfectly well. At length, on the morning of the 23rd of April, came an alarming telegram; we started off by the very next train, and Fanny was with her sister the rest of that day and all the night. But Mary was then apparently unconscious, or at least unable to show consciousness. Her spirit departed about 11 o'clock on the morning of the 24th, quite calmly, without pain or struggle. Indeed, throughout her illness, she appears to have been free from suffering with the exception of languor and debility. I cannot now expatiate on the merits of the precious sister whom we thus lost. She was, indeed, a precious being, a noble creature of God, a beautiful and glorious soul, beautiful alike in her intellectual and her moral nature. Her personal beauty was remarkable, even to the end of

1873. her life,—indeed beyond it, for when we saw her in her coffin, some days after death, the face was still beautiful, and looked in its fixed placid serenity, like an exquisite waxen statue. It is a great comfort to reflect upon her bright and happy life, cloudless beyond the ordinary lot of mortality, terminated by a painless death. She was made to be admired by all who saw her, loved by all who knew her, and loved and admired more and more as one knew her better. I suppose there never was a more devoted, a more perfect wife in this world; all her time, all her thoughts, all her energies, all her brilliant talents, were devoted to the helping and taking care of her husband, especially in these last two years, when he has been infirm and nearly blind. The anxiety about him has certainly thrown a shadow over these last years of her life. She has, contrary to all expectation, been spared the pang of losing him. No one could possibly have anticipated that she would be taken first. Mary was buried on the 30th of April, beside her father, in the cemetery at Woking. The service was read by Charles Kingsley—I need not say impressively. The others present (besides ourselves) were, Henry Lyell, Leonard and Arthur, Mrs. Young (her cousin), Edward, Louis Mallet, Mr. Twisleton, Mr. Symonds, Professor Hughes, Miss Buckley, Mr. and Mrs. William Prescott.

Before Mary Lyell's funeral, we were startled by learning that George Napier (of whose illness we had heard nothing before we came to town) was in an almost hopeless state, and on the 5th of May he

died. Cecilia had hurried up from Wales on
hearing of his great danger, and she and William,
as well as his own daughter, were with him to the
last. A complication of disorders carried him off at
the age of only 56 :—indeed, he had been generally
in bad health for some years past, and dangerously
ill a few years ago.—I regret him very much. He
had been one of my firmest friends ever since the
time when we were together at the Cape in '38·;
there could not be a truer, steadier, or more hearty
friend than he had been to me and my wife. He
was a most kind-hearted, sweet-tempered, genial,
brave and honourable man : a very pleasant
companion, with a good heart and head. We
should have felt his death more severely, if it had
not come just when our minds and feelings were
absorbed by the loss of Mary.

We came down to Barton on the 17th of May,
and have been quiet and alone here ever since,
enjoying the beauty of the country at this season,
the birds and flowers, and much soothed by them.

LETTER.

Barton,
June 1st, '73.

My Dear Katharine,

We have now been rather more than a
fortnight at home, and I have greatly enjoyed, and
am much the better for, the repose and tranquility
of the country and the beauty of this season.

1873. Fanny also, I hope and believe, is much better than when we left London. The return to regular and settled occupations, under such circumstances, also does me a great deal of good.

Fanny has read me a great many interesting extracts that she has made from dear Mary's letters: the more I recal and the more I read, the more I am struck with the beauty both of her intellectual and of her moral nature.

Though the weather has not been on the whole very fine since we came home—at least there has been very generally a cold wind, the vegetation is as fine as I ever saw it here, the flowering trees and shrubs in the greatest beauty.—The rhododendrons are just coming into flower, much more abundantly than usual at Barton, which is probably owing (as Mr. Fish has remarked in the *Gardeners' Chronicle*), to the unusual wet of the past season. In the conservatory we have now in blossom, Begonia boliviensis, Allamanda nerifolia, Clerodendron Balfourianum (very singular and beautiful), Vinca rosea, many varieties of Gloxinia, Saxifraga pyramidalis, besides more common things. Within doors, I am studying Grasses, which are very difficult, the exotic and tropical ones that is; and what makes them the more difficult, is, that hardly any two authors agree in the terms they apply to the different parts of the flower. I have gone through the greatest part of Decaisne, and Hooker's great book: the illustrations are most beautiful and most useful, they are the chief advantage that it has over Lindley's Vegetable Kingdom: otherwise, it

appears to me, an edition of that work brought up 1873.
to the knowledge of the day would have done as
well.—But the illustrations are everything.

I am reading steadily the " Antiquity of Man,"
and I find the *new* parts in this edition, such as
those relating to the Brixham Cave, the Aurignac
Cave, and the Caves of the Dordogne, very clear
and instructive.

I am now also reading the " Life of Grote,"
which is written with great spirit, but I do not care
for his politics, which I am now just in the midst of,
and shall be glad when I get to his Greek
History.

The new part of Rose Kingsley's travels in *Good
Words* of this month, is very entertaining. I cannot
say I feel any wish to follow her example, even if I
were young, but it must be very interesting to look
back, from the midst of the tame hum-drum life of
England, upon the wonderful countries she has
seen, and the adventures she has gone through.

Pray do not think yourself bound to answer this
letter; I felt an inclination to write to you, and
wished you to know that I was thinking of you. I
hope and trust that you are taking care of yourself,
and are better than when we were in London.

Give my best love to Charles Lyell, and my
kindest regards to the Miss Lyells, and believe me
ever,

Your very affectionate brother,
CHARLES J. F. BUNBURY.

JOURNAL.

1873. A beautiful day.

The Barton Labourers' Club feast.—I made them a short address.

———

Warmer, but showery.

We went in pony carriage to the Shrub—enjoyed the beauty of the bluebells and the song of the nightingales.

———

Mrs. Rickards came to lunch and spent the afternoon with us.

———

We took a walk in the Park with the little dog Harold.

We began to revise the arrangement of books in my study.

———

Went to morning Church and received the Sacrament.

———

Since we have been at home, I have received two

delightful and admirable letters from Sarah Hervey : 1873.
the one from London, of the 20th May, the other
of the 28th, from Alfreton in Derbyshire, where she
and the rest of the family were staying on a visit to
her future brother-in-law, Mr. Morewood, the
betrothed of Patience Hervey. In this she gives
me most delightfully animated and picturesque
descriptions of Alfreton Hall, Winfield, Matlock,
Chatsworth and Haddon Hall. She writes
admirably.

It is a great pain to me, that, in consequence of
our grievous loss in April, which has made us avoid
all society, I am debarred from the enjoyment of
Sarah Hervey's company, which, every spring for
some years past, has been one of my great
pleasures. But her letters make me some amends.

Wednesday, 11th June.

To Stutton in our usual way, with our own horses
to Stowmarket and thence posting.

Monday, June 16th.

We returned the day before yesterday from a visit
of three days (or rather two days and a half) to our
dear friends the Millses, at Stutton. They were, as·
they always are to us, most cordial, kindly, and
affectionate ; and our visit seemed to give them
pleasure. Mr. Mills, now in his 82nd year, fragile
in health, as he has long been, and now nearly blind,
is very cheerful, and declares himself to be remarkably well. Mrs. Mills is always exceedingly pleasant,

1873. as well as a warm friend.   She is more than twenty
years younger than her husband, and has much
anxiety about him ; and her own health is also, I
fear, very delicate.

Stutton is in great beauty ; the delicate varied
tints of its foliage, and its rhododendrons and other
flowers, in full perfection.   It is the most lovely spot
I know in Suffolk.

In a drive with Mrs. Mills through those beautiful
woods, called Holbrook Gardens (belonging to Mr.
Berners), I was struck with the superb growth and
rich blossoming of the rhododendrons.

The vegetation in general there is very luxuriant.
The profuse blossoms of the Red Campion Lychnis
diurna, very ornamental ; the " hyacinths " (Scilla
nutans) just past.

I was very much pleased at finding that curious
and interesting orchid, the Birds'-nest (Neottia
Nidus avis), in a grove in the grounds of the Rectory.   There were a great many plants of it.   I found
also Cochlearia Anglica in flower on the margin of
the river, but for the salt marsh plants in general,
we were too early.

One morning at Stutton, a fine trout was brought
to the breakfast table—a fine large red-fleshed fish,
caught at Hintlesham, Colonel Anstruthers'.   This
is worth noting, because, till of late years, there
were no trout at all anywhere in Suffolk, and the
waters of the county were thought not adapted to
produce them.   Not many years ago, Col. Anstruther
tried the experiment of hatching and rearing trout,
after F. Buckland's method, at Hintlesham, and

establishing them in a brook there. He seems to 1873.
have succeeded perfectly, and Hintlesham now pro-
duces fine trout. On the 12th of June, the Rev.
John Hervey came to luncheon with us, and after it
we went with Mrs. Mills to Wolverstone and saw
the conservatory there.

———— ————

Wednesday, 18th June.

Dear Katharine and Rosamond arrived.

————

Thursday, June 19th.

Katharine tells me that she has received a
letter from her friend Lady Smith, the widow
of Sir James, who completed her hundredth
year on the 12th of May last. She is still in
good health, and in perfect possession of her mental
faculties, though her sight is failing. She is much
pleased, and with reason, at the Queen having lately
sent her a copy of her Highland Journals, with her
autograph in it. Lady Smith was able to write with
her own hand to thank the Queen.

———— ——

Monday, 23rd June.

Dear Minnie Napier arrived.

————

Tuesday, 24th June.

Four of the young Campbells, Annie, Finetta,
and two of the boys arrived.

Friday, 27th June.

Beautiful weather.   Hay harvest going on briskly.
We all went to Hardwick, which was in great
beauty.

Saturday, 28th June.

The young Campbells went away in the afternoon.   Arthur MacMurdo arrived, and late in the
afternoon his mother came.

====

LETTER.

Barton,
June 28th, 73.

My Dearest Cissy,

I am going to write a few lines to you this
time instead of to Henry, because I want to tell you
how exceedingly I admire his courage and cheerfulness through all this most painful and trying
time.   It is most admirable, and appears to me
indeed wonderful, that suffering so much for such a
long time as he has done, and with so much to
justify the most melancholy apprehensions, he can
show not only so much firmness and composure, but
even write and talk with cheerfulness.   It shows a
noble character.   I am afraid I should fail very
much if I were put to such a trial.   And you, dear
Cissy, what a blessing it is both to him and to
yourself that you are able to bear up with so much
firmness and composure, and to face all the painful
scenes without being overcome.   I thank God, I

think there is a gleam of comfort in the last 1873. intelligence that Henry sent me.

Fanny and I are both well, I am thankful to say.

With love to William and Emily and your Emily, and above all to Henry.

Ever your affectionate brother,
CHARLES J. F. BUNBURY.

JOURNAL.

Monday, 30th June.

Since our return from Stutton, we have had some few of our dearest friends staying with us :— Katharine and Rosamond from the 18th to 26th : Minnie from the 23rd to the present time (and she will, I hope, stay with us till we go to London) : Annie and three others of the young Campbells (poor Edward's children*) from the 24th to the 28th, and Susan MacMurdo arriving on the 28th.

We are again in great and painful anxiety about Henry.

Annie Campbell (aged 15, but appearing older) is a very interesting girl, and will, I think, a few years hence, when her mind has been expanded by society, and has gained more freedom, be quite charming. Personally, I think she will be beautiful. Finetta is a lovely merry little darling of six, full of spirit and glee.

Susan MacMurdo looks very well after the years she has spent in India. She was in great spirits while with us, and gave most entertaining sketches

---

* Sir E. Campbell lost his dear wife in October, '72.

1873. of people she had met with in India or on their homeward way through Palestine and Syria :—in particular, an interesting and striking account of Lady Ellenborough.

LETTER.

Barton
July 1st, '73.

My Dearest Cissy,

I have this day written to Drummond to place £100 to Henry's account. It is a great pleasure to me to be able to help you in this way on such an occasion, for I know that the matters connected with this sad illness must altogether be a very heavy drain upon your resources. I should certainly think that the best and healthiest plan for you would be to go and stay in some healthy country place within an easy distance of London. Why not try the *Oatlands Park Hotel*, at Wey-bridge?—A first-rate hotel (at least it was so in '69), in the midst of a beautiful open park,—light soil and fine air. However, there may be other places preferable which I do not know of:—only I think that any healthy and agreeable situation near London would be better than going back to Wales for the time.

As I wrote so lately, I will add no more at present, except warmest thanks from Fanny and myself for your most loving and touching letter.

Believe me ever,

Your very affectionate brother,

CHARLES J. F. BUNBURY.

JOURNAL.

We, Fanny, Minnie, and I, with the little dog 1873.
Harold, drove to Thurston and saw Mrs. Rickards
and her pretty garden.

_____

Friday, 4th.

Violent storms of rain and some thunder. Fanny
and Minnie went to Mildenhall, and returned late.
Examined Chiococca from Brazil, and made a note
of it.

_____

July 9th, Wednesday.

Weather beautiful and very warm these last three
or four days, and very favourable to the hay-harvest,
which is going on briskly. The greatest part of the
park is now cleared, the hay either on the waggons
or already carried; and it is hoped that one very
large stack at the stables will be completed by this
evening. Scott tells me that it is in very fine
condition.

Barton is in great beauty of foliage and flower;
though we have seen the best of it in June; for the
rich, heavy leafy masses of this season do not equal
in beauty the more delicate and varied foliage of the
earlier summer. But the roses and honeysuckles
are now in their glory, and the white lilies beginning.
The Indian horse-chesnut in beautiful blossom.

D

1873.   [During these first days of July, Sir Charles
was busy with the revision of books in his study,
and he read the novel of "Old Kensington," by
Miss Thackeray, to Mrs. Napier and me.—F. J. B.]

———

July 10th.

Up to London.

———

Friday, 11th July,
48, Eaton Place.

We drove and walked in Battersea Park, which is
very pretty and very well laid out : the "Sub-
Tropical garden" beautiful at this season, with the
small Palms and Bananas, Cannas, Azaleas, Bam-
boos and other plants of beautiful foliage, intermixed
with gay, bright flowers.

We dined with the William Napiers : a family
party, with the one addition of *their* friend General
Hardinge.

———

Saturday, 12th.

Henry, Cissy, and Emmie came to luncheon with
us.   We went to see Katharine, and walked with
her about the Zoological gardens.   Always very
interesting and agreeable to see.   The birds very
numerous and fine.   Hornbills, two species, wonderfully strange creatures, with their enormous beaks
and the helmet-like appendages on them ; these
beaks apparently so cumbrous, really very hollow
and light.   These birds very tame with the keeper,
who handled them and played with them and
showed them off, putting his hand within their
beaks, and making them catch fruits which he threw

to them. Their feet very strangely formed ; their
cry very loud, wild, and harsh.

Flamingoes,—numerous, of two or three species—
very flourishing and lively—their singular propor-
tions, wonderful length of neck and leg, and
strangely-shaped beaks, not less striking than the
delicate beauty of colour of their plumage, and
especially of their wings.— Frigate-bird—a small
one ; young, I suppose ; wonderful length of its
wings, which the keeper made it display. The
Abyssinian hornbill, a rare and extraordinary bird,
in general appearance much like an enormous raven,
with clumsy legs, a monstrous beak, and great blue
wattles, like those of some gallinaceous birds, under
its throat. Pigeons in great variety : the crowned
pigeon, a magnificent bird ; the bronze-winged from
Australia (chalcophaps), very beautiful; the Nicobar
pigeon, splendid in plumage and peculiar in form—
not like a pigeon at first sight—covered with long,
drooping feathers, like the hackles of a cock, entirely of a brilliant, glossy, metallic green.

Weaver birds (Ploceus and allied genera), several
species, some of them very brilliant in plumage ;
some of their curious nests constructed on branches
very high up in the aviary. The keeper gave one
of the nests to Fanny, most curiously compacted of
interwoven grass.

The Nicobar Pigeon, *(Caloenas Nicobarica)*, very
well described in Sir W. Jardine's Naturalists
Library.

Katharine and her husband, Rosamond and
Arthur dined with us.

A visit to-day from Charles Lyell and his two sisters. I was struck by the apparent improvement in Charles Lyell's health : he not only looks better than he did last year, but his voice is in a surprising degree stronger and clearer. He talked for some time with much animation on his favourite geological topics, and discussed some difficult questions in a way which showed perfect clearness of intellect.

Edward dined with us :—very agreeable—many amusing stories.

One of Edward's stories :—

It is said that the present Shah of Persia is by no means a strict observer of the law of Mahomet against wine, and that he sometimes drinks a good deal. One evening, during his stay at Berlin, it is said, it was evident that he had drunk too much. Lord Odo Russell, on this being remarked to him, replied by the French proverb :—"*la nuit, tous les chats (Shahs) sont gris.*"

—————

July 18th.

My dear wife's 59th birthday. May God bless her.

We had the great and unexpected pleasure of a visit from Mrs. Kingsley and Rose. I had not seen Mrs. Kingsley since February, '69, and my delight at this meeting was great. She was not looking well, I am sorry to say, in fact she has had much ill-health : she is as charming as ever.

Montagu and Susan MacMurdo, their daughter 1873.
Emily ("Mimie") and Edward dined with us: very
agreeable. Montagu, as well as his wife, is in high
force, not looking at all the worse for India, and he
talks very well.

He told us of a remarkable military pass near
Beirout, in Syria, where the hills (spurs of the
Lebanon), approach so near to the shore as to leave
only room for a road. Inscriptions on the rocks
here in 4 different characters and 4 different ages,—
Assyrian, Egyptian, Greek and Roman.

July 19th.

Patience Hervey was married to Mr. Morewood
the day before yesterday, at Wells. Sarah has
written to us a very pleasant description of the
wedding, which seems to have been everything that
could be wished.

Drove with Fanny: we called on Mrs. Greig, saw
her; we looked for a short time into the Royal
Academy.

Monday 21st.

A splendid day—very hot.

The sad news of the Bishop of Winchester's
strangely sudden death, by a fall from his horse. I
knew him but slightly, but he was a man to be
much regretted—a wise and good man, extremely
agreeable, with a remarkable variety of knowledge
and accomplishments.

1873.   Lord Westbury also is gone, but his death has for some time been expected. I am very sorry for our friend Mrs. Abraham, his daughter, who had a deep affection for her father.

Drove with Fanny: saw Charles Lyell, Henry, Cissy, Emmie and William Napier.

———— ——

Very hot.

Our dinner party :—Louis and Fanny Mallet, the MacMurdos, Minnie Powys, William Napier, Mr. Walrond and young Mr. Sidney Osborne,—eight besides ourselves—a very pleasant small party, all the members of it very agreeable. In the evening, Lady Head and her daughter and two Miss Freres, very nice girls.

———— ——

A farewell visit from Charles Lyell and his sister. Our second dinner party :—Lady Head and her daughter Amabel, the MacMurdos, General and Mrs. Eyre, Mr. Theed (the sculptor) and Edward.

———— ——

Very fine and hot. Received another of Sarah Hervey's admirable letters.

We looked into Holloway's the print shop, to see Mr. Storey's (the American sculptor's) statue of " Jerusalem," at present exhibited there. It is impressive, dignified and noble, but I do not like

the effect of the tinting of the marble : it has a 1873.
livid look, like dead flesh—not like living.

We drove to Dickinson's, to look at the photograph he was painting of Mary.

* * *

Henry and Cissy left town yesterday on their
return to Wales, and Charles Lyell with his sister
Marianne set out the same day for Zurich, in
Switzerland, whither he is going to confer with
Professor Heer.

Henry is, I hope and trust, somewhat improved
in health.

We drove to Kew, and spent an hour-and-half
most agreeably (the day being beautiful) in the
gardens which are in great beauty ; the turf, the
flowers, and the shade of the fine old trees, very
enjoyable. All is admirably arranged and managed.
In the great Palm house, some of the Palms,
Musaceae and other Monocotyledons, are quite
magnificent in foliage and bulk, and really give some
idea of a tropical forest scene. In the Tropical
Fern house, an amazing variety of Ferns flourishing
in the utmost beauty and vigour.

* * *

Read a good sermon by the Bishop of Carlisle
(Goodwin), on Prayer. We both called on Sally,
and I called on the Benthams.

* * *

Our third dinner party:—the MacMurdos again, Octavia and Wilhelmina Legge, Charles Hoare, Mr. Eddis and Edward—all very pleasant.

———

29th July.

We drove, with Octavia Legge and Mr. Sidney Osborne, to the Zoological Gardens, and spent between one and two hours there very agreeably, seeing many interesting animals. A Chimpanzee shown to us by the keeper in a small room off the monkey house, wonderfully tame and intelligent, shook hands most gently with all the party, and opened the door on being told by the keeper to do so. Guillemots, their droll movements and clamour, their greediness for fish. Pelicans, several species, very much alike, but differing in details of plumage, all equally greedy,—their diving for fish. One very large old Pelican, which had been here many years, very tame and familiar with the keeper, who handled it freely, pulling its pouch about and stretching it like glove-leather to show its extraordinary extensibility. Flamingoes — singular appearance when they all ran, flapping their beautiful carmine-coloured wings, with their extravagantly long necks erect. One of them, apparently a peculiar species, much smaller than the rest, and of a uniform palish red: in the others the general plumage pale flesh-coloured, the wings of a peculiarly beautiful vivid and delicate red. Ibis, the so-called *scarlet* (the colour appears to me rather like red), and the Sacred, this white, with the head and neck and tail black.

A visit from our old friend Mr. Smith of Combe Hurst, whom we had not seen for some time. Very agreeable as he always was. Speaking of the "Life of Grote," he seemed not much to admire the way in which the work was executed; he said that every one loved Mr. Grote who knew him; but he thought there was too much of Mrs. Grote in the memoir, and he seemed to think her by no means pleasing.

————

July 30th, Wednesday.

Very hot. Down to Barton. Arthur and Clement joined us at Cambridge.

All well at home, thank God.

————

Thursday, 31st July.

Went into Bury to the assizes, and was foreman of the grand jury. A remarkably light calendar : only five prisoners for trial, and of those only two serious cases. Only one of the cases was of West Suffolk, and in that the culprit was a tramp. The whole business of the assizes, civil and criminal, was got through in one day. I dined with the Judges, Bramwell and Cleasby.

As we came down from London, we saw the wheat harvest begun in several places ; before the end of last week it seems to have become general on the lighter soils around Bury ; and to-day, Scott

1873. tells me nearly all the farmers in *this* parish will have begun cutting their wheat.

On Friday, the 1st of August; we had heavy showers with fine warm weather between. On Saturday 2nd, I walked to look at the young trees.

On Sunday 3rd, we went to morning Church, and received the Communion.

We are enjoying the shade and beauty of home.

The Vicar of Mildenhall, Mr. Henry Phillips, died on the 29th July.

———

August 5th.

We returned to Eaton Place; Arthur MacMurdo with us. The harvest appeared to be general in the country through which we passed.

———

Wednesday, August 6th.

Lady Bell, the MacMurdos, William Napier and Edward dined with us.

———

August 7th.

Very hot. The Bishop of Ely has been appointed to the vacant see of Winchester; an excellent appointment, but we are very sorry to lose our good Bishop.

Looked in to Gustave Doré's Gallery in Bond Street —again admired his noble picture of "Christ leaving the Prætorium," His "Tenebres."—Jerusalem at the time of the miraculous darkness, with the three crosses seen on the height in the background, black

against a glaring break in the dark, lurid clouds— 1873.
very striking and powerful, rather *sensational.*
Andromeda excellent in the naturalistic style.

Then to the Jermyn Street Museum of Geology,
where I saw several specimens of South-African
diamonds, one or two of them imbedded in the
rock, and also numerous instructive specimens of
the minerals accompanying the diamonds, and of
the rocks from which they seemed to be derived ;
all these are new additions to the collection. Also
several models of large and remarkable diamonds
from the same country ; one of them much larger
than the Koh-i-noor.

Visited the Leopold Powyses. Mr. Walrond
dined with us—very pleasant.

August 8th, Friday.

The Miss Richardsons, and the MacMurdos dined
with us.

August, 9th, Saturday.

We went to see the MacMurdos at Rose Bank,
and spent the afternoon with them.

We dined alone.

Monday, August 11th.

We dined with the Douglas Galtons—alone with
them—very pleasant.

Tuesday, August 12th.

Arrangements for sending Arthur MacMurdo and
his sister Louisa to Felixstowe.

We returne   yesterday from Tunbridge Wells,
whither we went on the 12th, and where we spent a
week very agreeably.

We were at the Calverly hotel, which is admirably
well situated, and though our room was small the
view from its windows was beautiful.

Tunbridge Wells made nearly as favourable an
impression on me, as the only time I was there
before—in '41.

It is a pretty little town, and as most of the
houses which have been built of late years are
scattered on the surrounding hills amidst trees and
gardens, they do not spoil the view.   The Common,
I was glad to see, remains in its former state,
covered with heath, furze, and fern, and wild grasses
and affording a most agreeable walk.   The
surrounding country is beautiful, and we enjoyed
several very agreeable drives.

The 13th, we went to see Hever Castle, eleven or
twelve miles from Tunbridge Wells; interesting as
the residence of Anne Boleyn in her maiden days,
and the place where Henry VIII. courted her; more
interesting for these associations than for anything
actually to be *seen*.

The drive thither is through a beautiful country,
rich, varied and smiling, hill and dale, pasture,
woodland and arable, agreeably intermixed;
abundance of hop-grounds; the villages and
scattered farm houses and cottages, thoroughly
rural in style. It is characteristically English
landscape of the best style.

Hever Castle lies low, at the bottom of a valley, 1873. amidst rather wet meadows, watered by the very small river Eden, which runs into the Medway. It is in style intermediate between castle and dwelling house; something near the same stage as Hurstmonceaux, but by no means as grand. The large massy square, castellated gate-house is much the most striking and imposing part of the building.

There is a double moat, the outer one now filled up with weeds; the inner very broad and full, fed by the Eden. The interior of the house occupied by a farmer's family, and not to be seen.

The Church on the hill above the Castle contains the tomb of Sir Thomas Boleyn, Anne's father.

The 14th was another beautiful day. Penshurst, which we saw on the 14th, is extremely interesting. The way thither, the same by which we returned from Hever. The House (like most of these grand old houses) lies low in a broadish valley through which flows the Medway, between gently sloping, well-wooded hills. The house is of various dates and various styles of architecture; its effect dignified and noble.

The first court one enters, reminds me strongly of an old college. The old hall called the Baron's hall, of Edward III.'s time, very grand, and especially striking as an example of the domestic arrangements of great houses in the old time; gloomy, arched, timber roof of vast height; stone, (brick?) floor; hearth in the centre, of the simplest construction, with iron bars to support the logs; rude, massy tables along the sides. Arms and

1873. armour of time of civil war on the walls.   The arrangements—of the dais at the end of the hall, the passage across from court to court at the other end, the relative positions of the buttery, kitchen, &c., reminded me much of Trinity College, Cambridge.   Staircase from one end of the dais of the hall to the upper apartments.   A fine gallery.

Very numerous family portraits, of very various merits, but many of them interesting from their subjects; in several I recognized the originals of prints in Lodge's work.   Sir Philip Sidney, Algernon Sidney, (two repetitions), Algernon Sidney and his two brothers as boys (very characteristic), Dorothy Sidney.

We returned from Penshurst by way of Bidborough, and stopped to look at the Church, which is nothing remarkable in itself, but in a very fine situation commanding a most extensive view over the beautiful country.

August 15th, we three (for Minnie had joined us) drove by way of Lamberhurst to Bayham Abbey, and back by Frant.   Lamberhurst, a pretty village on the slope and near the bottom of a steep hill.

Bayham Abbey is in Lord Camden's beautiful park; the situation not unlike that of Waverley—on low flat meadow land at the bottom of a valley, beside a slow stream, overlooked by the hill on which the new house is building.

The monastic remains which are very carefully and neatly kept, are not extensive nor grand, but remarkably graceful and pretty.

A beautiful drive from Bayham to Frant,

through Lord Camden's woods, over a bold, hilly 1873.
country ; the ground under the trees and along the
roadside richly clothed with flowering heath.
Frant green is on the top of a high hill, commanding an exceedingly fine and extensive view.

There is something about the remains of Bayham
Abbey, which gives a peculiar impression of graceful
and dignified repose. (The second stanza of
Keble's poem, in "The Christian Year"), on the
Monday in Whitsun week, seems to me peculiarly
expressive of the feelings suggested by this scene.

All this country is exceedingly pleasing and
smiling, varied in surface and richly wooded. I did
not, however, observe the same luxuriance of
growth, especially of wild vegetation in the hedges,
as in Somersetshire, especially in the country below
the Mendips, near Wells.

We were struck with the beauty and apparently
good construction of the cottages on the Penshurst
estate.

Hop grounds, a very characteristic feature of this
Tunbridge country, very beautiful when seen near,
and at this season, when the plant is in the full
luxuriance of its growth, and the full beauty of its
graceful wreaths.

August 16th, we first visited the high rocks, about
3 miles from Tunbridge Wells. These I had
seen in '41 : they are certainly remarkable,
especially for the depth, narrowness, and intricacy
of the vertical fissures or chasms which divide the
masses of rock, and some of which are of very considerable horizontal extent. These large abrupt

1873. rock-masses, overshadowed by and intermingled with trees and bushes, are very picturesque. The sandstone composing these rocks is fine-grained, soft, and almost crumbly, so that one wonders at its forming such bold and abrupt masses.

There may be some good botanizing, earlier in the summer, in the thickets about the high rocks, and in '41 I found some mosses there, but the weather now was much too dry for them. The Hymenophyllum which takes its name from Tunbridge, is said to be now quite extinct here : in '41 I found it in very small quantities on the high rocks.

Thence we drove to Buckhurst Park, and through Lord Delaware's beautiful woods to Withyham. Superb trees in Buckhurst woods, especially beeches and chesnuts. Withyham Church on a hill, adjoining the park. Curious monument in white marble, in the chancel, to Richard, Earl of Dorset, and his son, who both died in 1677, the boy reclining on a large tomb, his father and mother, both in full dress of the period, kneeling on either side: the whole very elaborate, cumbrous and heavy, without beauty.

A fine monument by Flaxman, to a Duke of Dorset, who was killed by a fall from his horse in Ireland.

Thence we drove through a beautiful country to Groombridge, a remarkably pretty and picturesque village, scattered on the side of a steep hill, amidst much wood, and returned to Tunbridge Wells by Langton Green, along high ground.

The 17th, Sunday, we visited Mr. Hamilton, 1873. the clergyman at Frant, whose house is delightfully situated.

August 19th, in returning to London, we stopped for several hours at Seven Oakes, for the purpose of seeing Knole park and house.

Knole* :—A grand old house of the Elizabethan style, grey sombre and venerable—appears to cover a great deal of ground. The interior specially remarkable for stately galleries of great length ; but I saw it too hastily to describe it. I do not know that it can be described better, especially as to the furniture, — than in Horace Walpole's letter to R. Bentley, p. 262 in the edition of 1798. —Great multitudes of pictures, many apparently copies of celebrated Italian ones—many interesting portraits.

The park of Knole is of very great extent (12 miles round, it is said), richly wooded and parts very beautiful; — noble timber. Above all, the largest beech tree, by far, that I ever saw,—really a gigantic tree, the trunk immense and the principal branches as large as the trunks of ordinary trees. Its foliage is thin, and altogether it has evidently seen its best days ; yet is still far from complete decay.

Also an oak, which has evidently been a very great tree, but is now quite a ruin ; not entirely dead however, for it still bears some leaves.

If this be "the Knole beech," mentioned in *Loudon's Arboretum*, p. 1977, as one of the largest

---

* What is the correct spelling of this name? It is *Knowle* in Horace Walpole; *Knole* in the Ordnance Map.

1873. trees of the kind in Britain, the diameter of the trunk is stated as 8ft. 4in., that of the head 352ft, and the height 85ft.

Another beech growing in the same park is mentioned by Loudon as 106ft. high, and 123ft. in the diameter of its head; and is said to grow in "pure sand."

The park is certainly on lower greensand, but the soil is probably not without some marl or clayey mixture.

There are at Knole very large and fine sycamores near the house.

While we were staying at the Calverley hotel at Tunbridge Wells, I renewed acquaintance with a lady whom I had not seen (I think) since 1837:— Mrs. Ward, formerly Catherine Houlton. I found her wonderfully little changed from what I remembered; still very handsome, and very friendly and cordial in her recognition of me. Her husband, Mr. Arthur Ward, is a very gentlemanlike man.

———— .

Wednesday, 20th August.

Weather stormy, wet and cold. Fanny spent the morning with Katharine. I went to join her there in afternoon, and found Frank there, just returned from South America. Minnie dined with us.

————

Thursday, 21st April.

Visits from Frank Lyell and Mrs. MacMurdo. Minnie dined with us.

————

Down to Barton. Mimi (Emily) MacMurdo with us. Arrived safe, and found all well at home. Thank God.

———

Sunday, 24th August.

We went to morning Church. A great thunder storm at night.

———

Monday, 25th August.

Very fine hot weather. Settling ourselves at home. Looked into "Life of Heber," and looked through some of Walpole's Letters for remarks on Knole, etc.

———

Tuesday, 26th.

A hot morning; a very heavy storm in afternoon. Looked over report of valuation of Mildenhall Vicarage.

Read part of Dampier's Voyage.

Arthur and his sister Louisa returned from Felix-towe.

———

Wednesday, 27th August.

Another talk with Scott about Mildenhall Vicarage lands. Drove with Fanny; we visited Mrs. Rickards.

———

Saturday, 30th.

Mr. and Mrs. Abraham came to luncheon. Minnie Boileau arrived. I went to a garden party at the Philipses.

Friday, 5th September.

Lady Hoste and Mrs. Burroughes came to luncheon and to see our pictures.

Took a drive with Minnie Boileau.

- - - - -

Sunday, 7th.

We went to morning Church and received the
Sacrament.

- - - - -

Monday, 8th September.

The children's party and merry-making on the
lawn.

A young Parrakeet* *(Calopsitta Novac Hollandiac)*
was hatched here on the 11th July, and is now
nearly fledged : weak in its legs, but otherwise
a promising bird.

Fanny has now three Parrakeets (male, female and
young) of this species ; a pair of the beautiful little
grass Parrakeet *Melopsittacus unditlatus.* (I follow the
names by which they are called in the Zoological
Gardens). A pair of Bulfinches, three Java
Sparrows and one common Sparrow.

- - - - -

September 8th.

The weather for more than a week past has been
very wet and cold, which has delayed the conclusion
of the harvest; but this day, Scott tells me it is
expected to be finished in this parish.

---

* 27th September. The poor little young Parrakeet died this day. It had
never been strong.

The wheat has been got up in very good condition 1873. but the barley, it is feared, has suffered from the cold rain.

- - -

September 9th.

Minnie Boileau went away, having been with us since the 30th of August. It was the first time we had seen her since her great misfortune in the death of her sister Theresa. Her visit has been a great pleasure to us. She is a really charming and admirable young woman; Mr. Walrond arrived with one of his boys.

- - -

September 10th, Wednesday.

Mr. Abraham (now an Honorary Canon of Ely) and his brother the Suffragan Bishop of Lichfield (formerly Bishop of Wellington in new Zealand) dined with us. The Bishop is a remarkably agreeable and interesting man : handsome and dignified in aspect, of very courteous and pleasing manners, with a great variety of knowledge and accomplishments, which makes his conversation very interesting.

- - -

Thursday, 11th September.

A fine day.

We went—Fanny and Mrs. Byrne in one carriage Mr. Walrond and I in another—to Ixworth, and visited Lady Hoste and Mr. Green. Lady Susan Milbank and her sister Lady Charlotte dined with us.

Friday, 12th.

Very fine.

A pleasant walk with Mr. Walrond. Lady
Cullum and two Miss Floods, Mrs. and Miss
Horton dined with us.

_____

September 15th.

Mr. Walrond went away. His stay here has
been exceedingly pleasant to us. He is indeed
delightful. I know few men who *suit* me better.

We were first attracted to him through his lovely
and charming wife, but he is much to be liked and
valued for his own sake. Though an official man,
he is remarkably modest, quiet and gentle in his
manners, extremely well informed ; his conversation
full of matter and very agreeable, without the least
display or the least tendency to excess in talk. All
his sentiments and ways of thinking, appear to me
those of a good, refined and honourable man.

_____

September 16th.

A large new book-case having been put up in my
inner study, I was busy much of this day, in
arranging part of my Natural History books in it ;—
they had quite overflowed the previous accom-
modation.

_____

Wednesday, 17th.

Attended a meeting at the Guildhall at Bury, to
resolve on Address to the Bishop, — Maitland
Wilson (High Sheriff) in the chair.

Dear Arthur MacMurdo and his two sisters went away—he to return to school, they to their home ; Henry Walrond (whom his father had left for a few days under our care) went with them. They have been very happy and joyous together, and very amusing : I miss their merry voices.

Emily MacMurdo who is nearly grown up (she is 16), is an uncommonly agreeable and interesting girl, and will grow, I think, into a charming woman. She is very pretty now, and will, I think, be prettier.

September 19th.

Received a letter from George Bentham, asking for information about my uncle Henry Fox, from whose large botanical collections I had sent many specimens to Sir William Hooker. Answered him, sending all the particulars I knew.

Scott brought me a very favourable report of the Mildenhall rent audit.

Saturday, 20th.

Mr. Theed and his son arrived.

The two Miss Floods, Lady Cullum's nieces, dined with us.

Monday, 22nd.

Mr. Percy Smith dined with us.

LETTER.

Barton
September 22nd, 73.

1873.   My Dear Katharine,

Many thanks for your kind letter; I am glad to hear that the rainy weather has not entirely prevented you from botanizing, and you have been in a rich country for Ferns, and no doubt for Mosses also. I often look at the plants which I collected at Kinnordy, and think of the old days. The other day I gathered the Fontinalis (not a common Moss hereabouts) in the very same spot where I first found it in 1825, when I was a very young bryologist, a long time ago, and when my mother was the confidant of my delight in the discovery.—Plants which one has one's self gathered and ascertained, have, according to my experience, a great power of *sentimental* association; there are some good remarks on this subject in Sir James Smith's Memoirs or in his tour, I am not sure which.

I am very glad that Charles Lyell has been able to do so much geological work to his satisfaction in Switzerland, though it must have been a great disappointment to Miss Lyell not to be able to see the Alps. What a blessing his devotion to geology has been to him under his great affliction, and what an advantage to have, not only good sisters and sisters-in-law, but an active young friend like Mr. Hughes to act as *eyes* for him.

I shall be very glad to see Charles and his sister 1873.
here ; it is unlucky that we had already, not
knowing his plans, made up our party for this week.
I found his geological letters very difficult to
understand, not being very conversant with the
particular questions which he wanted to clear up.

Your Forfar photographer amused me very much,
and his remarks on the hand-writing* appeared to
me decidedly shrewd, though sufficiently uncom-
plimentary. As you are used to *that lady's* hand-
writing, her journals will have given you ample
information as to all our doings, and I am afraid I
have but little to tell you.

I have been rather desultory—partly because I
am bothered about the vacant living of Mildenhall.
I wonder any one can ever wish for patronage.

We had delightful visits from Minnie Boileau and
Mr. Walrond, both especial favourites of mine,
and the MacMurdo children also were very pleasant
inmates of the house : I miss the merry voices and
rapid feet which so long made it echo. Emily (or
Mimi), the eldest of them, is a really charming
girl for her age. Mr. Theed, the sculptor, and his
son are with us now—both pleasant.

I look forward with great anticipations to next
month, in the course of which we hope to see here,
not only Charles Lyell, but also later on Mr.
Kingsley and his daughter and the Arthur Herveys.
Sometimes I feel almost afraid to expect anything
so pleasant, I feel a vague dread of something

* My hand-writing which was much criticised by my friends and
relations.—(F. J. B.)

1873. coming to baffle my hopes: but on the whole I
believe Kingsley is right,—that he is happy who
expects everything, "for he enjoys every thing once
at least, and if it falls out true, twice also."

It is very curious what you mention about the
Cystopteris without indusium; I think I have read
of other instances which seem to show that the
indusium in Ferns is not of such constant importance as is generally supposed.

Ever your affectionate brother,
CHARLES J. F. BUNBURY.

JOURNAL.

Tuesday, 23rd.

The Barton Rent Audit, very satisfactory—
luncheon afterwards with my tenants and Mr. Percy
Smith. Lady Charlotte Osborne and Mrs. Horton
came to luncheon with Fanny.

Wednesday, 24th.

Lord Charles and Lady Harriett Hervey and
Miss Louisa Hervey, Sir Frederick and Lady Grey,
Lady and Miss Frere, Cecil and Clement, Minnie
Napier and John Herbert arrived.

Thursday, 25th.

Mrs. Byrne and the two Mr. Theeds went away.

Cecil was ordered off suddenly by the military
authorities.

A pleasant walk with Lord Charles Hervey.

Lady Hoste and Mr. Greene, Lady Ellice and Miss Ellice and Mr. Abraham dined with us.

———   —

September 29th.

A pleasant party in the house this last week :—
Lord Charles and Lady Harriet Hervey and their daughter Louisa, Sir Frederick and Lady Grey, Lady and Miss Frere (Sir Bartle unfortunately could not come), Minnie Napier and her brother John, and my nephew Clement.

The Charles Herveys left us on Saturday the 27th, Lady and Miss Frere this morning. Lord Charles Hervey I have described in previous years, and I continue to think him very agreeable and thoroughly estimable. He is most earnest and conscientious, kind-hearted and amiable, with much of the charm of his brother Lord Arthur. His wife, Lady Harriet, is an excellent and very pleasant person, with much intellectual activity and cultivation.

Miss Frere very pleasant.

———   —

Monday, 29th.

Sir Edward and Lady Greathead, Mr. and Mrs. Wilmot Horton arrived.

Mr. and Lady Louisa Wells, the Maitland Wilsons, Sir Charles and Lady Ellice and Miss Berners dined with us.

———

The last 7 or 8 days of this month have been
extremely fine :—that beautiful serene autumn
weather which we often have in this part of
England, which is as enjoyable as any weather that
I know.   Some days the sky has been perfectly
cloudless.   The colouring of the foliage, especially
of the exotic trees, is becoming very beautiful.

Sir Frederick Grey thinks it probable that the
information obtained by Sir S. Baker, respecting
the lakes Sanganyika and Albert Nyanza may be
correct,—that there may really be a communication
between the two, not constantly and regularly, but
in the season of floods, the one overflowing into the
other.

The Hortons dined with us this day.

————

October 1st, Wednesday.

The two Miss Floods came to luncheon.

Finished reading the Commentary (Lord Arthur
Hervey's) on the two books of Samuel.   It contains
much interesting information, and clears up some
difficult passages, but I wish I could say that he
had dealt satisfactorily with two *moral* difficulties
which occur towards the end of the book.   I wish
he had expressed indignation at the abominably
cruel sacrifice of Saul's seven grandsons (II. Samuel,
ch. 21st) to satisfy the Gibeonites.   It is to me
impossible to believe that this atrocity was com-
manded or sanctioned by the Almighty.   The
sacrifice of men who (as far as appears) were them-

selves unoffending, in order to put a stop to a 1873. famine, appears to me a superstition worthy of barbarians.—Neither can I believe that the divine anger was excited against David because he made an enumeration of the people, or that this " sin " was the cause of a pestilence, (see II. Samuel, last chapter.

October 2nd, Thursday.

A fine day and very warm—even hot.

October 4th.

The Frederick Greys left us. They are both of them agreeable and interesting, with active, vigorous, and well-stored minds : and it is pleasant to see a long-married couple so devoted to each other.

Charles Lyell and his sister Marianne arrived, also our dear Rose Kingsley, who has not been at Barton since February, 1869, though we have seen her in London since the beginning of this year. She is as charming as ever.

Sunday, 5th.

We went to morning Church and received the Sacrament.

Tuesday, 7th.

Fanny's new goats arrived.
Examined fruit of Magnolia acuminata.

1873.   Rose Kingsley tells me, that her brother
Maurice (who has been surveying in the wildest
parts of North America, in the Western States
and Mexico), one day found himself on a "rattle-
snake ledge," such as is described in "Elsie
Venner." He had climbed on to a ledge of rocks
on the side of a mountain without observing
anything particular, when hearing a rattle, he looked
round, saw a snake and struck at it. Immediately
the whole rock was alive with rattlesnakes, of all
sizes coming forth from every crevice, swarming
and hissing and rattling around him, it was like
the serpent scene in Gustave Doré's Dante. He
took a flying leap over the rocks, and ran down the
mountain side at his utmost speed, regardless of
any other dangers; he declares (Rose says) that he
never leaped so far in his life before, nor ran so fast.

Rose, herself, never saw a rattlesnake while she
was in Colorado, because it was the winter season;
but since then, Maurice has killed one very near to
where they lived.

October 8th.

Charles Lyell and his sister went back to London.
He is in tolerable health, except that he is nearly
blind, which is a sad misfortune: but otherwise his
health appears to be decidedly better than it was a
year ago, and better than we could have expected it
to be after such an illness as he had the year before
last, and after the dreadful loss that he suffered
last April. His sister Marianne, who now lives with
him, is an intelligent and excellent woman, and

seems to take very good care of him. They are 1873.
just returned from Switzerland, where they have
been making a tour, accompanied by Professor
Hughes of Cambridge, and where Lyell conferred
with most of the principal geologists, and gained a
good deal of information that was interesting and
valuable to him. His devotion to geology is, if
possible, more intense than ever. The questions to
which his attention seems now to be most especially
directed are—the formation of lake-basins by the
excavating action of glaciers, or otherwise, the
transport of granitic blocks from the Alps to the
Jura, whether by floating ice or by continuous
glaciers, (he, I think, inclines to the latter theory).

Charles Kingsley arrived; and thoroughly
delighted I was to see him again in my house. We
had a most agreeable evening of conversation. He
seemed well in health and his talk is as animated
and as rich in matter as ever.

The last time he was here was in February, 1869.

Thursday, 9th.

Spent the morning very pleasantly in looking
over dried plants with Kingsley—very good talk also
at breakfast and luncheon.

Mr. Robeson arrived.

LETTER.

Barton, Bury St. Edmund's,
October 9th, 1873.

My Dear Katharine,
    I am very much obliged to you for the fine

1873. plant of Aspidium Lonchites which you sent me,
and which is a valuable addition to our set of out-
door Ferns.   Allan has just put it under a frame
for the present for precaution, but I hardly
suppose, that a plant which grows naturally at the
elevation of as much as 6000 feet on the Alps, can
require permanent protection here against anything
but the parching winds of March.

Fanny has got some new pets: a couple of
Thibet goats, just now given her by Mrs. Evans
Lombe, very pretty animals and excessively tame
and familiar; they are kept in an iron-fenced
enclosure on the lawn, with the old hut or " chalet "
for their head-quarters, and I hope they will never
stray, for they would make awful havoc among the
choice shrubs.

It makes me melancholy to see dear Charles
Lyell so blind, but in other respects he seems to me
quite as well as one could possibly have expected,
and decidedly better than he was a year ago.   He
sometimes, indeed, appears nervous, but when he
was alone with me talking geology, he spoke
perfectly clearly, and without any apparent hesita-
tion or difficulty.   The *uncertainty* of his walk
seemed to me to be owing to his blindness, and not
to feebleness.   His sister appeared to take excellent
care of him.   I was very sorry they could not stay a
day longer, so as to meet Charles Kingsley.   *He* is
now here as well as his daughter Rose—both quite
delightful ; I revel in talk with them about tropical
American botany, and snakes and insects and birds,
and a variety of other subjects.

Edward expects to be in London about the 11th 1873. or 12th—he seems to have made a successful and pleasant tour in the Savoy and Dauphiné Alps.

Our trees are now coming to be in great beauty of autumn colouring.

Believe me ever,
Your affectionate brother,
CHARLES J. F. BUNBURY.

JOURNAL.

Friday, 10th.

Fanny and Rose went to Mildenhall and returned to dinner. Mr. Robeson also went thither with Scott, to see the place.

Saturday, 11th.

Mr. Robeson went away.

A very pleasant morning of herbarium study with Kingsley.

Montagu and Susan MacMurdo and Mimi, Maurice Kingsley, General and Mrs. and Miss Eyre and Mr Pryor arrived.

Sunday 12th.

We all went to morning Church. Charles Kingsley preached in our Church a noble sermon* on part of the 104th Psalm. It was in fact a mag-

* The sermon which now stands as the 16th, in his collected volume of
Westminster sermons.

F

1873. nificent discourse on Natural Theology; inculcating in the grandest and most eloquent manner, the great truth that God is the Author and Creator of all things, and that what we call Nature and the Powers of Nature, and the Laws of Nature, are only modes in which He manifests His Will.

I have been revelling in conversation with Charles Kingsley and with Rose about the botany and natural history of Tropical America. His ardour, his remarkable talent and habit of observation and his great power of expression, make all that he says about tropical nature very valuable. His daughter is scarcely inferior to him in either talent. She is indeed a very remarkable young woman, and a most charming one.

Monday, 13th.

General and Mrs. Rumley and Frank Lyell arrived.

Had a pleasant walk with Susan MacMurdo.

Tuesday 14th.

A beautiful day.

A delightful walk with Rose Kingsley, and a good botanical morning with her father.

Sir Edward and Lady Greathead arrived, also Edward and Lord Augustus Hervey.

Wednesday, 15th.

Another good botanical morning with Kingsley— interesting talk with him. In afternoon a very agreeable walk with him and Maurice.

Mr. Lott arrived.

The Abrahams and Mr. and Miss Rodwell dined with us.

A favourable answer from Rev. Mr. Robeson.*

I consulted Kingsley on the difficult subject of prayer. He fully acknowledged the difficulties of the question, and spoke on it with great frankness and fairness, in a spirit of trustfulness and devotion, without a particle of austerity or cant, which was very comfortable to me. He spoke with great indignation of Tyndall, and some others, who have sought to deride those who attach a value to prayer. He concluded by saying that the intellectual difficulties concerning prayer, must in fact be solved by praying; that *so* only should we learn really to know the truth on the subject. " *Solvitur precando*," he said. His religion is to my mind both elevating and comforting.

Kingsley thinks very unfavourably of the present state and prospects of France, he says what can you expect, where the best hopes of order and good government rest on such a "jackanapes" as Thiers? He thinks that the ruin of France under this last Empire was brought about by the priests, and especially the Jesuits, working on the Emperor through the Empress, over whom they had gained absolute power. He has no doubts that the Emperor was urged on to the German war by the Jesuits.

Kingsley sets a great value on our aristocratic institutions; but he agreed with me that spendthrift

* Accepting the living of Mildenhall.

1873. and profligate noblemen (such as too many are)
are the cause of more mischief and danger to those
institutions than all the Odgers and Bradlaughs.
He would have a *censorial* power (as the ancient
Romans would have called it) residing in the House
of Lords itself, not only to expel unworthy members,
but to place them under control. I remarked what
a strange jumble of inconsistent ideas there is in
the 6th Æneid—some parts so grossly materialist,
others displaying a lofty Platonic Pantheism.
Kingsley entirely agreed with me, but thought that
there were not less glaring incongruities in Dante.
I dare say he is right.

Thursday, 16th.

A very beautiful day. Out all the morning with
Fanny, Charles Kingsley and Scott, selecting and
judging trees.

Friday, 17th.

A delightful morning spent in looking over
botanical drawing and specimens with Rose and
her father.

Took a walk with Lord Augustus Hervey.

Clement returned.

A remarkably agreeable party in our house since
the 11th. Besides the Kingsleys (father and
daughter and the son Maurice), we have had the
MacMurdos, and their pretty and clever daughter
Mimi, Sir Edward and Lady Greathead (for part of
the time) General and Mrs. Rumley, Lord Augustus

Hervey, Mr. Marlborough Pryor (part of the time), 1873. Edward and dear Minnie Napier.

Montagu MacMurdo is in great force and certainly is one of the most agreeable men I know, and of uncommon force of intellect and character. His wife also is very agreeable and wonderfully handsome for the mother of more than a dozen children.

Both are looking remarkably well after their years in India.

Kingsley said that the word *lawn* originally and properly means an open grassy space enclosed by woods. This was in answer to my remark that "*russet*" (as in Milton's "Allegro") seemed to me a singular epithet for lawns. He thought the epithet sufficiently descriptive of such *lawns* in the autumn, when they are clothed with the dry, withered stalks of the grass. He says that the blindness of West Indians in general to outward nature is something amazing. (Amusing anecdote in illustration of this) :—

"A West Indian gentleman returning from England to his Island, carrying with him as a treasure a plant of *Asclepias Curassavica* which he had bought, from an English nursery garden ; and finding to his astonishment that it was a perfect weed in his own and his neighbours' gardens."

Coral Snake :—Kingsley recognized this (from Brazil) in my drawing, as apparently the same that he knew in Trinidad. He said there was the same strange conflict of testimony, as to its venomous or non-venomous qualities in Trinidad, as I had found in Brazil ; a conflict which

1873. he did not know how to reconcile. Some he said, had resorted to the hypothesis that there are *two* species, so much alike as to be hardly distinguishable by any outward marks, but belonging, the one to a venomous and the other to a harmless group.

The common Maple.—Kingsley remarked that this tree grows much larger here, about Barton, than he has been used to see it elsewhere.

Small Pox :—Quinine in enormous doses, has been found (in Mauritius I think) to be the most effectual remedy against the most malignant forms of this disease (Kingsley).

Kingsley told me that he had strongly recommended the cultivation of the *Cork* tree on a large scale, to a friend who has an estate on the exposed west coast of Ireland; that nothing bears the *sea-gales* so well, and that in that soft, warm climate it would probably thrive so well as to be very profitable.

October 18th.

The Kingsleys, MacMurdos, and almost all our other guests went away.

Of Charles Kingsley, I must repeat what I wrote in my journal nearly thirteen years ago, that I feel it a great privilege to have the friendship of such a man, and to be able to receive him in my house. His conversation is wonderfully rich, various and interesting, his knowledge most extensive, both of books and men, and, what renders all these intellectual gifts thoroughly delightful, is, that I feel perfectly sure both of the excellence of his heart, 1873. and of his devotion to moral duty.

Of his daughter Rose, my opinion is even higher than it was: she is, indeed, an admirable and noble creature, and a charming one,—noble both in intellect and character. Her eager love of knowledge and devotion to all noble and elevating pursuits, do not give the least hardness to her character or manners, nor impair the charm of her natural gaiety and frankness. It would be difficult to praise her too highly.

I had a delightful walk with Rose on the 14th, and two or three times with her father, besides many interesting mornings spent in looking over my herbarium with him.—In fact, nearly all these days deserved very *white marks.*

Maurice Kingsley, the son, is an interesting young man, of very gentle and pleasing manners, and of much intelligence, he seems to have the family talent for observation.

Rose is not regularly handsome, but her expressive countenance is to me very fascinating.

I am not sure whether it was Kingsley or Sir Edward Greathead, who told *this* of the Prince Consort—that not long before his death he said to one of his friends:—" I know I am not popular in England, I never shall be, I have no *tact*, no German has."

Kingsley said that the Prince Consort was delightful when you were alone with him: but he was so shy or reserved, that the moment a third person joined the group, though it might be one of

1873. his particular friends, he immediately became cold and stiff.

Kingsley said, that roughness, and a disregard of the sensitiveness of others, are characteristic of the people of the *north* of England, just as they are of Germans:—that if there ever were a civil war between the people of the north, and those of the south of England, and the northern people were to conquer the south, they would make themselves odious just as the Germans did in France.

———

October 24th.

We returned the day before yesterday from Ely, where we had spent two days at the Palace, as the guests of our excellent Bishop and Mrs. Browne. It was on the occasion of a great festival in honour of the 1200th anniversary of the foundation of the Cathedral.   We went on the 20th, the Maitland Wilsons and Augustus and Lady Mary Phipps being in the same railway compartment with us: we reached the Palace in time to be present at the reading of farewell addresses to the Bishop from the several Archdeaconries of the Diocese; he replied with great feeling and excellent taste.

Next was an immense public "collation" or luncheon :—very bad fare, and many long speeches, most of which I heard but imperfectly.   I sat next to Mr. Beresford Hope.

In the evening a great "reception" at the Palace, an immense number of people.   Here I met Kingsley again, also Lord Charles Hervey, Professor Hughes, Mr. Lott and some others.

There was a pleasant party staying at the Palace, 1873. and among these we had the delight of meeting the Arthur Herveys, the Bishop, Lady Arthur and Sarah. There were also the Archbishop of Canterbury and Mrs. Tait, the Phippses, Mr. and Mrs. Hervey and Lord Crewe.

We attended afternoon service in the Cathedral— a grand "choral" service, in which a great number of choirs from all parts of the Diocese took part. It was a grand and striking pageant in that glorious Cathedral. The Bishop of Bath and Wells preached an excellent sermon.

The 22nd we returned home, by way of Thetford as we had gone. The same day a great many friends and acquaintances arrived—our party for the Bury Ball. Cissy and Emmy, Minnie Powys, Octavia and Mina Legge, Mr. and Mrs. Montgomerie, John Hervey, Mr. W. Hoare, Mr. Holmes, and two or three others arrived.

Lady Cullum dined with us.

———

Thursday, 23rd.

A delightful walk with Sarah Hervey and the rest of the young ladies. In the evening most of our party went to the Bury Ball : the Arthur Herveys, Cissy, Minnie, Fanny and I staying at home.

———

Friday, 24th.

Another delightful walk with Sarah Hervey and the young ladies. Our dance very successful— pretty ladies.

Saturday, 25th.

John Hervey, Major Grant, Mr. Holmes, Mr. W. Hoare and the Montgomeries went away. Mr. Walrond, Edward, Admiral Spencer and Lord John Hervey arrived.

LETTER.

Barton, Bury St. Edmund's.
October 28th, 73.

My Dear Katharine,

I was very sorry to hear of Miss Fanny Lyell's death, because I knew it would be a great sorrow to you and Charles and Harry : and besides, though it is so long since I saw her, I never can forget the great kindness that she and all the sisters showed us during the long time that we were detained at Kinnordy by my Fanny's illness.

I hope you are now enjoying as gloriously fine weather as we have here at present, after the desperate storms which prevailed during our visit to Ely. The colouring of the foliage here is very beautiful this autumn.

I have not at all been disappointed in my expectations of pleasure from the visits of the Kingsleys and the Arthur Herveys. Both have been thoroughly delightful, as much so, almost as I can imagine any thing on this earth.

I spent many agreeable mornings in conversation with Mr. Kingsley over my herbarium, or in walks with him, and learned much from him. Altogether, it has been a time to be remembered as long as I

retain my faculties, and for which I can never be 1873.
sufficiently grateful. In truth, I think there are
few things in this life for which one has more reason
to be thankful, than for such society as at once
satisfies one's taste, one's intellect, and one's moral
faculties.

I have begun to study a very important work on
fossil botany, which I have lately bought, Schimper's
"*Traité de Paléontologie Végétale*," three thick 8vo.
volumes, with an Atlas of plates. As far as I have
yet read, it is very judicious, and written with
French clearness, though the author, I believe, has
elected to remain at Strasbourg and to become a
subject of Germany.

Fanny is, I hope I may say, pretty well, and
though she has been the life and directing spirit and
organizer of all our society, and has worked
immensely, I think she looks well, and does not
appear to suffer from all her fatigue.

I hope your Frank will before long be able to find
some suitable employment. I should have said
before, that I was much pleased with him when he
was here. It is very difficult now-a-days to find a
good career for a young man who is a gentleman by
family, education and habits.

Pray give my love to your husband and
Rosamond, and any of your sons who may be with
you, and believe me,

Your affectionate brother,

CHARLES J. F. BUNBURY.

JOURNAL.

1873.   Another very beautiful day.

A long and pleasant walk with Fanny, Sarah, the Bishop, Minnie Powys, the Legge girls, Mr. Walrond, and John Hervey.

A visit from Lady Cullum.

Sir John Shelley arrived.

A most agreeable walk in the morning with Sarah, the Bishop, and Lady Octavia.

The Arthur Herveys left us ;—I need not say how very, *very* sorry I was to part with them.

Mr. and Mrs. Fazakerly arrived. Most of our party went off to the Thetford Ball.

The dear Legge girls went away,—also Admiral Spencer.

We walked about the grounds all the morning with Mr. and Mrs. Fazakerly, Minnie Powys, Sir John Shelley, and Emmie. Minnie Powys and I went to dine with the Wilsons—met the Arthur Herveys.

All our company went away, except Minnie

Napier, Cissy, Emily, and Clement—and all these 1873.
with Fanny went to luncheon at Stowlangtoft.

A visit from Captain Horton.

Wrote to Mr. Nicholl, sending him a cheque
for Lord Sandwich's rent.

Clement went away.

We two and Minnie dined with Lady Hoste and
Mr. Greene at Ixworth Abbey.

We dined with the Benyons at Culford—met
Lady Manners, Miss Newton, Mr. Rodwell, and
others.

Mrs. Walpole and Miss Greene came to luncheon.

The company we have had with us since our
return from Ely has been very agreeable. Those
three charming girls who have so often played
an important part in our autumn meetings—Minnie
Powys and Octavia and Wilhelmina Legge ; dear
Cissy and her daughter; Mr. and Mrs. Montgomerie ;
Mr. and Mrs. Fazakerly. (Mrs. Fazakerly is sister
of Minnie Powys) ; Mr. Walrond (part of the time) ;
Edward ; Admiral Spencer ; John Hervey and his
cousin Lord John ; Mr. Sancroft Holmes (a very
intelligent and agreeable young man ; Sir John
Shelley (very pleasant and intelligent) ; these, not
all at the same time.

1873.   November 7th.

Began on the 5th to read Herodotus in Greek.

- - - —

Saturday, 8th November.

Henry arrived, seemingly tolerably well.

═══

LETTER.

Barton,
November 8th, 1873.

My Dear Lyell,

I will with pleasure read over your chapter on the Vegetation of the Coal Period, and give you my opinion as to whether anything in it requires alteration. In the meantime I can send you a few notes. A passage which I extracted from *Nature*, last year, and which I believe to be quite correct, will give you an idea of the uncertainty of our knowledge respecting Coal plants :—

"Adolphe Brogniart and Dr. Dawson believe the "*Sigillaria* to be *Gymnospermous Exogens;* Carruthers "and Williamson (and Schimper) believe them "to be *Lepidodendroid Cryptogams.* Schimper, "Carruthers and Williamson regard the whole "of *Calamites* as Cryptogamens and allied to "Equisetum ; Brogniart, and perhaps Dawson, "think some of them are *Equisetais*, and some "Gymnospermous Exogens ; Carruthers and Binny "consider the fruits known as *Volkmannia Binnigi* "as the cones of Calamites ; Williamson disputes "this. Dawson thinks that *Annulariæ* and

" *Asterophyllites* are very distinct ; Carruthers thinks 1873.
" the contrary. Carruthers (and Schimper) believe
" that both *Asterophyllites* and *Annulariæ* are the
" foliage of *Calamites ;* Williamson that they are
" very different from Calamites."

Amidst all this conflict of opinions, it is indeed
difficult to draw up a concise statement (such as
you want for your book), which shall be satisfactory
or even fair to all parties. Almost the only points
on which all seem to be agreed are, as to the Ferns
(on which there is no room for doubt) and the
Lepidodendrons : these latter being undoubtedly
Lycopodiaceous, and Lepidostrobi being the spikes
of fructification.

For my part, I feel pretty well satisfied that the
Calamites were Cryptogamous and near to
Equisetum, and that the Sigillariæ were also
Cryptogamous, and allied (though perhaps not very
closely) to Lepidodendron.

It appears to me also very probable, though not
certain, that Asterophyllites were the leafy twigs of
Calamites.

In Equisetum, the verticillate leaves are united
into sheaths : in Calamites (if Schimper and
Carruthers are correct), the leaves are verticillate,
free, and not at all sheathing.

I know a good deal about Williamson's papers.
Five of them have in the last few years been referred
to me by the Royal Society, and I have gone
carefully through them, and reported upon them.
They are excessively elaborate, very lengthy, wordy
and tedious, but very careful, and full of curious and

1873. minute observations, supplying a great store of valuable materials for future systematists. Fortunately there are good clear abstracts of them in the Proceedings of the Society, and I will refer to these for information for you.

I am studying Schimper's great work, " Traité de Paléontologie Végétale," very good and instructive, though I differ from him on some points. He quotes me often ; I think he is rather too much inclined to assume that two fossils must be different species if they come from what are supposed to be different formations, by which method one is in danger of reasoning *in a circle.*

(November 9th). Having read over again your chapter 24, I think there is no occasion for alteration as far as relates to the Ferns and the Lycopodiaceæ. As to the Calamites, it might perhaps be well to mention that the specimens of them which are commonly found in coal mines, and preserved in collections (and such as is represented in your figure 460), are merely inorganic *casts* of the internal cavity of the stem, the ridges and furrows being impressions of the inner surface of the woody zone. The little tubercles commonly seen near the articulations (as in your fig. 460), appear to be the scars of broken bundles of vessels. You might also mention, that the Calamites differed from living Equisetums chiefly in the more complex and more highly organized structure of the woody zone, resembling in appearance the exogenous structure. (Observe that the question whether this structure was *really and essentially* exogenous, is one of those

on which Williamson is most at variance with   1873.
Carruthers and others).   With respect to *Stigmaria*,
I do not see that you notice that *Lepidodendrons*
have been found to have the same kind of roots
(Stigmaria) as the *Sigillariae*.   I forget whether it
was Dr. Dawson or Mr. Richard Brown who
ascertained this fact, but it appears to me
important, as helping to prove the affinity of
Lepi. and Sig.   (The names are so very long I
must shorten them to save space).   Prof.
Williamson, I think, considers Sigillariae as allied
to Lepidodendrons, but with a more highly
developed wood.   As to Antholithes, I cannot help
still doubting its belonging to a flowering plant, or
at least one of a high order.   The last information
that I have picked up concerning it, is this extract
from *Nature*, Feb. 2, '71 :—

"At a meeting of the Royal Phys. Soc. of
" Edinburgh, Mr. Peach exhibited specimens of
" *Antholithes Pitcairniae*, with *its fruit* (Cardiocarpon)
" attached, from the coal-field near Falkirk."

Now *Cardiocarpon* has generally been considered
as belonging either to something allied to Lepidodendrons or to Cycadeæ : Schimper places it
among the " fructus incertae sedis."   I should think
the Antholithes may perhaps belong to something
like a Cycas, but hardly to anything of a higher
order.   I am very glad of the information about
Huxley's startling theory concerning the structure of
coal.   It surprised me very much when I first met
with it in my friend Kingsley's " Town Geology,"
and on inquiry I found that he had derived it from

G

1873. Huxley.   I suppose it may be true of some
peculiar kinds of coal, and that Huxley generalized
rashly.

This packet is quite large enough, so I will close
now and write again in a few days if I find anything
more to communicate.   I am very glad to hear
that Miss Lyell has returned to you safe and well
(in these days of railway accidents, one is always
glad to hear of a friend having safely arrived).

Pray give my kind remembrances to her.   We
are both well, and my brother Henry who is with us
just now, seems, for him, tolerably well,

Ever yours affectionately,

CHARLES J. F. BUNBURY.

JOURNAL.

Sunday, 9th.

An   admirable   sermon   from   Mr.   P.   Smith   at
morning Church.

Monday, 10th November.

Charles Lyell wrote to me two days ago, asking
me for some observations on the Flora of the Coal
Period, for the 24th chapter of his new edition of
the "Students' Elements.".

I sent off to-day a long and careful letter, giving
him all the information (from Williamson, Schimper
and others) that I could compress into the space.

Interrupted by Mr. Bain about the hospital.

November 11th. 1873.

My dear brother and sister Henry and Cecilia, their daughter Emily and dear Minnie went away.

Henry seems, for him, in tolerable health. Minnie is quite a sister to us, and a delightful companion.

We went to Ickworth to luncheon with Lord and Lady Bristol, and met the Arthur Herveys.

———

Wednesday, 12th.

Received another geological letter from Charles Lyell.

———

LETTERS.

Barton,
November 13th.

My Dear Lyell,

Your books have not yet arrived, so I can say nothing about Antholithes or Pothocites; but, while waiting for them, I may answer your question about Asterophyllites. What Carruthers says on this latter subject (in his paper in the *Journal of Botany*, Dec., 1867, of which he sent me a separate copy), is this:—" The whorled leaves described under the name of *Asterophyllites*, have been frequently found joined to stems of *Calamites*, putting beyond all doubt that they are the foliage of at least some species of this genus." He shows that the spikes or cones which he calls *Volkmannia Binneyi*, and which he elaborately describes in that

G 2

1873. paper, are the fructification of Asterophyllites, and *therefore* (according to his view) of Calamites.

What Schimper says is merely, " On est aujourd'hui généralement d'accord que les Asterophyllites sont les rameaux feuillés des Calamites." On the other hand, Williamson, in the latest published notes of his on the subject that I have seen (*Royal Society Proceedings*, June 1872), says that specimens he had lately examined—"finally settle the point that Asterophyllites is *not* the branch and foliage of a Calamite, but an altogether distinct type of vegetation, having an internal organization peculiarly its own." For my own part I do not feel able to decide between the two opinions, *both* professedly resting on positive observed facts. The only explanation I can suggest, is, that there may be plants of different families agreeing in the supposed characters of Asterophyllites, and that some of them may have belonged to Calamites, and others *not*. I cannot say that I think much of Dr. Dawson's argument founded on the midrib; the leaves of different kinds of Calamites may have varied in this character. But, indeed, what are the other leaves "known to belong" to Calamites ? Let me remark that the phrase *Monocotyledonous angiosperms* (p. 413, l. 1, of your "Students' Elements," ed. 1), is superfluous, for no known Monocotyledons are other than *angiospermous*. Whether the Coniferæ, etc., are truly Gymnospermous, is at present a subject of keen controversy, as you may see in Bentham's last Linnean Address ; but it is most likely that Robert Brown was right, and at any rate, that they are in some 1873. measure a natural group seems proved by their geological history. But the phrase should be *gymnospermous* or *angiospermous Dicotyledons :* for the higher and more comprehensive class is the Dicotyledonous one, which includes the Gymnospermous and angiospermous groups or subclasses. (6 p.m.)—As your books have not arrived, I will send this as it is, and write again after I have studied the documents you are sending me concerning Antholithus, etc.

Ever yours affectionately,

CHARLES J. F. BUNBURY.

JOURNAL.

November 14th.

Mr. Bowyer, inspector of workhouse schools, came on the 11th, and went away to-day. He is a very clever and remarkably well read man. I began to read aloud to Fanny in the evening, Help's "Spanish Conquests in America."

LETTER.

Barton,
November 15th.

My Dear Lyell,

I am sorry I can give you no help as to the subject referred to in your last letter, for I have had no practice in examining slices of coal or other things

1873. under such high magnifying powers, and Mr. Newton's sketches are rather unintelligible to me. But I can explain the "*macrospores*" mentioned in Huxley's note.   In many recent Lycopodiaceæ, there are two different sorts of "capsules" or "spore *cases*" (sporangia), bivalve reniform ones, containing minute dust-like spores, and others which are 3-lobed or 4-lobed, and 3 or 4-valved, and which contain just as many larger spores as they have valves.   These larger spores are called *macrospores*, the others microspores.   You will find all this clearly represented in Le Maout Decaisne and Hooker's great book (which I think you have) at page 913,—also in Hooker's " British Ferns," pl. 52, (Lycopodium selaginoides, which is the only British species that has the two sorts of spore cases).   Williamson's conjecture that the supposed spore cases may be macrospores, seems probable enough.   Your books have just arrived, but I cannot enter upon them this evening.

Ever yours affectionately,
CHARLES J. F. BUNBURY.

JOURNAL.

Monday, 17th.

Dear Kate and Charles Hoare arrived, a very pleasant evening with them.

Tuesday, November 18th.

A week ago I received a long letter from Professor Parlatore of Florence, inviting me in the most courteous manner and with many flattering 1873. expressions, to attend the Botanical Congress to be held at Florence in May next, and to act on the jury of the Horticultural Exhibition. The invitation was a flattering one, but for various reasons we have thought it best to decline it.

LETTER.

Barton,
November 18th, 1873.

My Dear Lyell,

Carruthers' article on Anthrolithes, in the *Geological Magazine* (which I am very much obliged to you for lending to me) is very instructive and important, and gives me quite a new idea of the nature of these fossils. The specimens he describes and figures, show clearly, I think, that what had been taken for flowers are the pedicels of fruits, with clustered bracts accompanying them, The name *Antholithes* is therefore evidently quite inappropriate, and I think he is quite right in snbstituting that of *Cardiocarpon*. It strikes me that, if it be not too late, it would be highly desirable to connect the few lines relating to this fossil in your "Students' Elements" (p. 412), and to give a new wood-cut, copied from Carruthers' *Cardiocarpon Lindleyi* (see p. 55 of the paper I have been speaking of) in place of your fig. 472. You will see in the paper aforesaid, that Carruthers no longer considers it to be an "angiospermous Dicotyledon allied to Orobanche," but on the whole is "inclined

1873. to consider Cardiocarpon as a *Gymnosperm* of an extinct type." (p. 56, near the bottom of the page). You will see that he compares it to the fruit of some of the Yew groups (Taxineæ). At page 58 he adds "It is possible that *Cardiocarpon* may have been the fruit of the Taxineæus *Dadoxylon*, and that the *Trigonocarpon* may have been the seed of a large form of Cardiocarpon." On the whole I do not think that the so-called *Antholithes* can be quoted as evidence of the existence either of Monocotyledons, or of angiospermous Dicotyledons in the coal formation. With respect to the *Pothocites*. I do not find anything in the *Geological Magazine* which gives me more information about it than I find in your book. The drawing certainly has in a general way, the look of an Aroideous spadex, but I am not familiar with the structure of the Aroideæ (except the European ones) and can give no decided opinions upon it. I have also read with much satisfaction, Dr. Dawson's paper in the *American Journal of Science*, and I think he has very well made out his case.

The MS. addition which you have prepared to insert in your book, on the structure of coal, seems to give a very fair and clear *résumé* of what has been ascertained on the subject. By the way, do you want this paper returned to you? I rather suppose not, from the way in which you mentioned it in your letter. The books I will send back by railway as soon as I can,

Ever yours affectionately,

CHARLES J. F. BUNBURY.

JOURNAL.

We two, with Charles Hoare, dined with the 1873. Wilsons at Stowlangtoft, Kate not being well enough to go out; we met Barnardiston, Harcourt Powell, and Henry Wilson.

———

November 20th.

The death of old Mr. Phillips* the Vicar of Mildenhall, at the end of July, imposed upon me the task of nominating a new clergyman; a heavy charge, and a serious responsibility.

After making many enquiries, I offered it to a friend of Mr. Walrond, Mr. Robeson, a Balliol man, who has hitherto had the small living of Forthampton, near Tewkesbury; and he accepted it. Mr. and Mrs. Robeson came here on a visit to us, the day before yesterday, and left us yesterday afternoon to stay with Mr. Lott at Barton Mills. We have a pleasant impression of both of them.

———

Friday, 21st.

Fanny went to Mildenhall, and returned to dinner.

═══════

* He had been Vicar of Mildenhall, mostly non-resident, since 1817.

LETTER.

1873. My Dear Lyell,

Thanks for giving me an opportunity of correcting a passage in my former letter, which certainly was wanting in clearness. I aimed at brevity, and became obscure. Certainly I did not mean that *Dicotyledon* ought to be substituted for Monocotyledon in the passage relating to Pothocites. What I meant was, that in one or two other places (such as p. 402, "Students' Elements,") where you use the expression *Dicotyledonous angiosperms*, it would be better to say *angiospermous* Dicotyledons : Dicotyledons being the more comprehensive term, which includes the other.—At least such is my view, and I think Hooker would agree with me. Lindley, I know, considers the *"Gymnogens"* (gymnosperm Dicotyledons) as one of the primary groups of Phanerogamous plants : but I think the general opinion of botanists is against him. All the alteration I should propose as relating to *Pothocites* would be, in l. 1, p. 413, to put *Monocotyledons* in the place of Monocotyledonous *angiosperms*. I shall be glad to read over bye-and-bye, the passages inserted or corrected as you propose to send them.

I am glad your sisters are now all with you. I hope it will be a comfort to all to be together. Pray give my kind remembrances to them.

Ever yours affectionately,

CHARLES J. F. BUNBURY.

JOURNAL.

Dear Kate (Hervey) Hoare and her husband 1873. came to us on Monday, the 17th, and went away on Friday, the 21st, a very short visit, but very pleasant to us.

Kate is as charming as ever, and looking very pretty, though, I fear she has not strong health. Charles Hoare very pleasant, and I fully believe very estimable. We are now quite alone, and I hope that in what remains of this year, I may get through some good reading.

========

LETTER.

Barton,
November 23rd, 1873.

My Dear Lyell,

I think the wood-cut of Carruthers' *Cardiocarpon anomalum* will do very well for your " Elements," and give a sufficient idea of the general structure, though *Cardiocarpon Lindleyi* is perhaps more distinctly characteristic as to the fruit. That paper of Carruthers' appeared to me so important that I sent to Mitchell for a separate copy of the number of the *Geological Magazine*, which contains it. I have got it, but I am much obliged to Mr. Carruthers for his offer of a copy.

As to the form of the word — the Greek *karpos* is used with various terminations in botanical

1873. names. I think the termination in *us* is the most frequently used—as *Artocarpus*, Gyrocarpus, Elæo-carpus, Echinocarpus—but there are examples to support both *on* and *um*. I think I would use Cardiocarp*on*, as it is already published, and I do not know of any objection to it.

I had overlooked the paper from Dr. Dawson, but I will send it back to you in a day or two, after looking at the passages you have marked.

Ever yours affectionately,

CHARLES J. F. BUNBURY.

JOURNAL.

Tuesday, 25th.

We dined with James Bevan at Bury.

Wednesday, 26th.

Disagreeable information respecting dilapidations at Mildenhall.

Saturday, 29th.

We called on the Percy Smiths, and went with Mr. Smith to look at the new coloured window in the Church.

Monday, December 1st.

Signed the form of Presentation to the Living of Mildenhall for Mr. Robeson.

Lady Susan Milbank, Lady Ellice, Mrs. Wilson
and Agnes came to luncheon.

———

LETTER.

Barton,
December 2nd.

My Dear Lyell,

I am very glad to help you as far as I can
in the matter of the fossil plants, but I feel that I
am, myself, on anything but sure ground in giving
opinions upon them. It is indeed excessively
difficult, amidst such a *vortex* of conflicting opinions,
to draw up a statement which shall be at once clear,
concise, and fair to all the combatants. The worst
of it is that there seem to be not merely differences
of opinion and inference, but sometimes differences
of actual observation as to matters of fact.
Professor Williamson evidently wished to make you
commit yourself to a positive statement in favour of
his views, but if I were you I would keep to my
neutrality. I am not sure that your statement, in
the MS. extract, which you enclosed, is quite clear.
I have written on a separate piece of paper what
occurs to me as what I might put in its place, and
what appears to me sufficient, as you cannot, within
such narrow limits, enter into particulars of the
internal structure. But of course this is only a
suggestion, and you may have very good reasons for
keeping to what you have written. As to

1873. *Volkmannia*, you will find figures of one species in Carruthers' paper on the Fruit of Calamites (*Journal of Botany*, Dec. 1867), and of another, the original species (*V. Polystacha*) in Sternberg's old book, " Flora du Monde Primitiff," (tab. 51). I can send you Carruthers' paper if you like, I have it bound up with a number of other things. I should not think it is necessary for you to give a figure of any Volkmannia, for they are rare fossils, and not characteristic of the formation like the Asterophyllites, Sphenophyllum, Lepidostrobi, etc. It appears, from that last memoir of Dr. Dawson which you sent me a little while ago, that he has seen, or believes he has, Calamites *in situ* with their branches and leaves attached, and that *these* are *not* Asterophyllites. Has he published any detailed description of such specimens? I begin to think that there is a preponderance of testimony *against* the identification of Asterophyllites with Calamites, but I do not think it is as yet conclusive, and it appears much better in your " Elements," to notice briefly the various opinions.

(December 3rd). Really, the more I read about these questions, the more puzzling they appear. The different authorities seem to contradict not only one another, but themselves sometimes! I thought that Williamson considered both *Asterophyllites* and *Annularia* as quite distinct from Calamites, but in the paper which you last sent me (at p. 34), he says :—" That plants of the Annularia type constituted the delicate aerial foliage of Calamites, appears to be established by so many

independent observations, as to leave little room for 1873. doubting the conclusion. It does not follow from this, however, that *all* the *Asterophyllites* and *Annularia* are Calamitean." This was in 1871. In '72, in his latest paper that I can find in the *Royal Society Proceedings*, he says nothing about *Annularia*, but says most positively that *Asterophyllites* has nothing to do with Calamites, but has a very peculiar organization, much more like that of Lycopodiaceæ, and that in this peculiar organization his *Volkmannia Dawsoni* agrees remarkably with Asterophyllites. His figures of this V. Dawsoni give no idea of its external appearance, but I suppose that in *that* respect it was not entirely unlike the original *Volkmannia* of Sternberg—that is, that it was something like a catkin, or the spike of an Equisetum lengthened out.

I now think it best to cancel what I had written as an attempt at a summary of the question. But if you wish I will make another attempt, a little less hastily in a few days. In the meantime I send you back your papers.

I find, in Schimper's "Paléontologie," some excellent remarks on fossil *Proteaceæ*, but that is a subject which you are not concerned with at present.

I am very sorry to hear that Miss Lyell continues in such a bad state of health, I hope she will soon get well.

Ever yours affectionately
Charles J. F. Bunbury.

Barton.
                                    December 4th, 73.

My Dear Katharine,

It is a long time since I have written to you, but I have (as you know) been keeping up such an active correspondence with Charles Lyell about fossil plants, that you have at least known how I was employed; and I have besides been plagued and worried about some Mildenhall business. But I really ought to have written to you sooner to congratulate you on Leonard's triumphs, which it gave me very great pleasure to hear of. It is most highly honourable to him to have gained such distinguished places in such very severe and difficult examinations, and is very gratifying to all who are connected with him, most of all of course to you.— But he must have worked terribly hard for these honours. I trust that his health will not have been seriously hurt. I am very glad he is gone to Italy, and I hope he will amuse himself and enjoy a little of the *dolce far niente* in that delightful country, and that it will do him a great deal of good.

I am deep in Schimper's "Paléontologie Végétale," which is very good indeed, and have had plenty to do in consulting books to answer Charles' queries. How wonderfully energetic and active-minded he is!

In these short dark days I cannot do much in my Museum. Fanny and I have just finished cataloguing the Natural History books which I have lately arranged in my new bookcase. In the evening I read to her Help's " Spanish Conquest of

America," which is *painfully* interesting. I am also 1873.
reading (to myself) Sismondi's "Républiques
Italiennes." I have looked through the fine (fine
in the sense of showy) new book on "Holland
House,"—entertaining, but rather flippant and
flimsy, I think. I shall like very much to read
Mrs. Somerville's "Autobiography," when it is
published.

We are going to spend the greater part of next
week at Mildenhall (where I hope we shall not be
perished with cold), to make arrangements for Mr.
and Mrs. Robeson, as well as to look at the
plantations, etc. That house is haunted by the
ghosts of past joys and sorrows.

Pray remember me most kindly to Miss Lyell and
to both her sisters.

Believe me ever,

Your affectionate brother,

CHARLES J. F. BUNBURY.

Barton,

December 5th, 73.

My Dear Lyell,

The objection I should make to the statement in your last bit of MS. concerning Asterophyllites, etc., is, that it gives prominence only to
Dawson's argument (from the midrib of the leaves),
to which I should myself be inclined to attach less
weight than to Williamson's argument from the
internal structure.

I enclose a note of what appears to me, if I
rightly understand the matter, to be a brief and fair

1873. statement of Williamson's view, and I shall be very glad if you find it useful to you.

I only mentioned slightly Schimper's opinion on the *Proteaceæ*, because I supposed that the part of your book to which they belong had been long since done with. What he says on the subject is, in substance this:—That both parties have gone too far. That there is no reason for denying the existence of the remains of true Proteaceæ in Europe in the upper Cretaceous and lower Tertiary formations. But that, on the other hand, he is satisfied that a considerable number of the fossil leaves referred to that family by various authors belong really to quite different families, and that a much greater number are altogether doubtful. In particular, he says that the leaves of *Myricaceæ (Myrica* and *Comptonia)* cannot be distinguished with any certainty, in mere impressions, from those of Proteaceæ. He says that some naturalists (he names particularly Unger and Ettingshausen), took up a pre-conceived idea that an "Australian vegetable, physiognomy," predominated in Europe through certain periods, and proceeded on this assumption to refer to Australian genera—a great many fossil plants which might just as well, or better, have been referred to genera or families belonging to other parts of the world.—So far Schimper. I am always extremely glad if I can help to throw any light for you upon any of these questions.

Ever yours affectionately,

CHARLES J. F. BUNBURY.

NOTE.— " Botanists are not yet agreed as to

whether the *Asterophyllites*, as species of which is 1873.
represented in the annexed fig. 461, was the foliage
of the Calamites or not. Prof. Williamson, from
the microscopical examination of the internal
structure of many well preserved specimens, has
come to the conclusion that the *Asterophyllites* did
not belong even to the same natural order with the
Calamites, but formed a distinct group, likewise
Cryptogamous, but more nearly related to the
Lycopodiums than to the *Equisetums*. (Then might
follow the opinions of Dawson, Carruthers and
Schimper, as in your former MS.)

On the subject of *Annularia* and *Sphenophyllum*, it
might be added that there still are differences of
opinion as to whether these forms are, or are not
closely allied to Asterophyllites."

JOURNAL.

December 7th.

I believe I have hitherto forgotten to mention
Fanny's new pets:—two Thibet goats, given her
by Mrs. Evans Lombe.

They arrived on the 7th of October, and have
thriven well, being kept in the stables in bad
weather and at night, and at other times in a small
enclosure in the pleasure-ground. They are very
handsome animals, and ridiculously tame and familiar. They are, I suppose, what are called
Cashmere-shawl goats, in the Zoological Society.
One of them is brown, the other black and white.

1873.   My correspondence with Charles Lyell on the
Carboniferous Flora, has been brisk. The day
before yesterday I wrote to him my 8th letter (since
the 7th of November), on the subject, and this
morning I had another letter from him, saying that
what I had written in my last on the subject of
Asterophyllites was exactly what he wanted, and he
would adopt it *verbatim* in his new edition. Since
the 21st of last month we have been quite alone,
with plenty of occupation and very happy.

I have gone on steadily reading to Fanny in the
evenings, Help's "Spanish Conquest of America ;"
—very interesting. We have worked together at
filling up our library catalogue, by entering in it all
the books recently arranged in the new bookcase in
my room. I have also read the greater part of the
1st volume of Sismondi's "Républiques Italiennes."

My reading of Herodotus has, I must confess,
rather flagged, and I have for the present *stuck* at
chapter 60 of the 1st book.

I have lately run through the showy new book on
"Holland House," by the Princess Mary of Licht-
enstein. There is a very favourable review* of this
book in the October number of the *Quarterly :*—the
" article " certainly better worth reading than the
book. (The review* by the way appeared nearly a
month before the work was published).

December 8th, Monday.

We went to Mildenhall.

* This review was written I believe by Mr. Hayward.—F. J. B.

We went out first to the Nursery Cottage, saw Julia Betts—then with George Betts to see the young plantations—satisfactory.

A long visit from Mr. Lott.

A very fine day.—sharp frost. We went out again with G. Betts; first to Curll's plantation, then to see two new cottages at Beck Row.

Fine and very cold.
We visited the boys' school.

Intensely cold—fog in morning and rime frost.

Mr. Lott and Miss Bucke came to luncheon.

News that Marshal Bazaine is found guilty and condemned to death and to military degradation.

The sentence is probably just; but yet I hope it will be mitigated. (It has been commuted by the President to twenty years "seclusion;" and he is to be spared the *ceremony* of degradation. Dec. 13th).

Since we came hither (the 8th) I have read the political article in the October number of the *Quarterly*—known to be by Lord Salisbury — the "Programme of the Radicals." It is very clever, very brilliant, very powerful, and I believe there is a great deal of truth in his conclusions.

1873.   December 13th.

We returned home from Mildenhall, whither we had gone partly to look at young plantations and new cottages, and partly that Fanny might get the old Manor House into good order for Mr. and Mrs. Robeson (the new clergyman and his wife), to whom we have lent it, as there is no proper Vicarage house at Mildenhall.

The young plantations appear to flourish, and to make satisfactory progress.

A new piece of ground has been fenced and planted every successive year for several successive years, and the special object is to grow good larches, which are found to be more useful and profitable, especially in that district, than any other tree.

The new cottages—two at Beck Row and one on the Brandon road, are very satisfactory.

Monday, 15th December.

Archdeacon Chapman came to luncheon.

Tuesday, 16th.

Walked with Fanny and the little dog Harold.

Began to arrange some old notes on botany, and wrote tickets for the new books.

Wednesday, 17th.

Dear Arthur MacMurdo arrived.

Thursday, 18th December.

Received a charming letter from Sarah Hervey,

and a very pleasant one (with a present of a book), 1873. from Clement.

Wrote some of my Arboretum Notes. Mrs. Rickards and Mrs. Wilmot* came to luncheon.

LETTER.

Barton,
December 23rd, 73.

My Dear Katharine,

I congratulate you on Leonard's engagement to deliver a course of Lectures, which I am very glad indeed to hear of, as it is not only (as you say) a good start in the career of a *Professor*, but is an excellent opportunity of diffusing knowledge and of communicating his opinions and views on those subjects which he has studied so well.—I have not the least doubt he will go through the course very successfully. It is only a pity that he should have to hurry back from Italy.

I am very glad, too, that you have such a gratifying character of Arthur† from Mr. La Touche, and no doubt well deserved. It is very delightful to see all your children growing up to repay you so well for the careful education and the earnest thought and love you have bestowed on them.

I have nearly read through J. S. Mills's "Autobiography," a most singular and very interesting book. Almost all his opinions are

---

* Mrs. Wilmot was a Darwin; a very clever and interesting woman, sister-in-law to Mrs. Rickards.
† Arthur Lyell.

1873. utterly antagonistic to mine—absolutely repugnant to me, and yet there is something interesting and almost attractive in his character. But if I had known him in his life-time, I do not suppose we should have sympathized on any one subject except botany! No, this is going too far; I agree very much with him also in his zeal for individual liberty, in which he was at variance with some of the extreme Liberals of the present day. A good deal of what he says about Logic and " Psychology " I do not understand. I remember, when I saw a fine photograph of him in Mrs. Kingsley's book, I was struck by the unhappy expression of his face, and the general tone of his book is quite in accordance with this.

(December 24). We are very near now to the end of the year, a time which always makes one thoughtful. To you, I know, it has been a year of deep sorrow, of sadness and anxiety: and to me, also, it has been marked by much sorrow, for, besides that grievous loss which is common to us with you, I lost a dear and valuable old friend, a most steady, genial and hearty friend in the death of George Napier: and besides, we have repeatedly been in painful anxiety about Henry. But yet, I am far from looking back on the past year as on the whole a time of gloom and sadness; I feel that I have had, and have, very much of happiness to be thankful for, and I trust that this is your case also, and that there are yet many bright years in store for you. As I advance in life, and as I feel that I have little or nothing to wish for (in *this* life) for myself, I feel

more and more how much happiness may be 1873.
derived from one's hopes and anticipations for one's
young friends, from one's interest in their careers,
and from the sight of their enjoyment of life.

I am afraid I am growing *prosy*, but I think you
will understand me.

With all my heart I wish you and all your family
a happy Christmas and New Year, and many of
them.

I trust that Charles Lyell is not over-working
himself.

Ever your very affectionate brother,

CHARLES J. F. BUNBURY.

The death of Agassiz is a great loss to science,
and I should think unexpected.

JOURNAL.

Christmas Day.

We went to morning Church.
The servants' supper party.

Friday, 26th.

Henry sent us, to read, a very entertaining letter
from his son Henry, from Carthagena.

We entered some new books in our catalogue.

Tuesday, 30th December.

Wrote to Sarah Hervey, and sent her Mrs.
Somerville's memoir as a present : sent cards to
Octavia and Wilhelmina Legge.

Visit from Lady Cullum.

A beautiful sunny day.   Walk with Fanny.

The servants' dance and merry making.

The year which is now coming to an end, has been, to us, one of great and startling sorrows, but also one of much happiness and many blessings. The unexpected death of dear Mary Lyell*—so utterly unexpected till within a few days of its occurrence, was a heavy blow, a deep grief, an irreparable loss to both of us, of course most so to Fanny. I shall only add, to what I wrote at the time, that though the lapse of time abates the poignancy of grief, memory does but bring out her merits in brighter colours the more it dwells upon her.

Almost at the same time we lost a most staunch, true-hearted, genial and excellent friend by the death of George Napier, a most kindly, pleasant and honourable man : a friend, to me, of long standing, and specially associated with the memory of pleasant days at the Cape. These are indeed two heavy losses.

My cousin, General Fox, died in the spring, a very good and pleasant man : and in January, Adam Sedgwick passed away, serene and happy, in extreme old age, leaving a venerable and honoured name.

Henry has, in the course of this year, sold Abergwynant very advantageously, and bought a little place in the fine healthy country, near

---

* I think there can hardly be a doubt that she was one of the first victims of the terrible outbreak of typhoid fever in Marylebone, caused by polluted milk.

F. J. B.

Wokingham, which Cecilia and Emily are very much pleased with.

Charles Lyell, whose state of health caused us so much anxiety for the last two or three years, is yet in a condition far from robust: we cannot but feel that a fatal attack at any time would not be surprising. Yet he is certainly better than he was last year. All his faculties and powers are concentrated on his favourite geological pursuits, at which he works with wonderful energy and perseverance, and his letters on that subject show that (in that direction at least), his mind is as clear and vigorous as ever.

Who could possible anticipate at the beginning of this year, that Mary, who seemed so full of life, would be taken away, and her invalid husband be left?

Among the many blessings for which I feel that I have cause to be deeply grateful, after the love and help and comfort of my admirable wife, I feel especially thankful for many valuable and delightful friendships which I am so fortunate as to enjoy, among the young as well as among those of my own generation.—Such friends as the Arthur Herveys, the Kingsleys, Mr. Walrond, Minnie Powys—to say nothing of the Napiers and MacMurdos, who are more nearly connected with us, are gifts for which one cannot be too thankful. Though we lost our regular London dissipation, we have enjoyed some society of the highest order both for pleasure and improvement. I have seldom passed a more delightful month, in this respect, than this last

1873. October.   The recollection of such a time is a
lasting source of happiness.

My correspondence of letters with Sarah Hervey
continues to be actively kept up, and affords me
very great enjoyment.

My scientific correspondence with Charles Lyell,
which has been brisk in the last two months, has
been a vigorous and useful exercise to my mental
faculties, at the same time that it has enabled me to
give some help to my friend, and to contribute
something through him to the advancement of
knowledge.

A large number of eminent or conspicuous men
have passed away in the course of this year :—the
Emperor Napoleon the Third, Manzoni, Lord
Lytton, Sedgwick, Liebig, John Stuart Mill, Dr.
Lushington, Bishop Wilberforce, Landseer, Count
Strzelecki, Agassiz, Sir Henry Holland.

BOOKS READ, 1873.

Grote's History of Greece.   Chapter 78-90 (to end
of vol. 11).

Froude.   English in Ireland.   Vol. 1.

Sismondi.   Républiques Italiennes.   Chapter 1-12.

Helps.   Spanish Conquests in America.   Vol. 1
and greater part of vol. 2.

Memoir of Grote.   1 vol.

Autobiography of John Stuart Mill.   1 vol.

Mrs. Somerville's Memoirs. 1 vol. 1873.

Geikie. Scenery and Geology of Scotland. 1 vol.

Ramsay. Physical Geology and Geography of Great Britain. 1 vol.

Lyell. Antiquity of Man. Edition 4. 1 vol.

Wyville Thomson. Depths of the Sea. 1 vol.

Decaisne and J. D. Hooker. General System of Botany (not read through, of course, but much studied).

Schimper. Traité de Paléontologie Végétale. 2 vols. (A great part studied).

Hayward. Biographical and Critical Essays. 2nd series. 2 vols.

Houghton (Lord). Monographs. 1 vol.

Holland House. By Princess Mary of Liechtenstein. 2 vols.

A Princess of Thule. By W. Black.

1874.

JOURNAL.

1874. Received one of Sarah Hervey's delightful letters, with excellent remarks on Mills's Autobiography.

Very fine—no frost.

Sent some money to Mr. Babington, for charities at Cockfield.

We finished catalogue of new books in my room.

Arthur and his sister Louisa arrived.

The labourers' supper party.

We went to morning Church and received the Communion.

A fine day with sharp frost. The children's party —a large gathering and much merriment. Alice Praed* stayed here.

Clement arrived.

Mr. Robeson was yesterday both *instituted* and *inducted* as Vicar of Mildenhall. Fanny went over to see the ceremony ; but the bad weather and fear of a cold kept me at home. She took Arthur and Louisa with her and returned at 6.

* Now Mrs. Holden.

LETTER.

Barton,
January 9th, 1874.

My Dear Susan,

I thank you very heartily for your kind 1874.
expressions towards me in your letter of the 2nd,
and I can sincerely assure you that I fully return
the feelings of friendship and affection which you
express towards me. The past year did indeed
bring heavy and unexpected sorrow to us, and you
are quite right in saying that I shared in the grief of
that irreparable loss which befel us. I loved Mary
very much, and good reason I had to do so, and
very, very much I admired her. And though time
abates the poignancy of grief, yet memory, the more
it dwells on her brilliant and beautiful and noble
qualities, does but bring them out in brighter
colours. I often think how agreeable and charming
she was the last (or almost the last) time I ever saw
her alive, on the 30th March, just before we left
London, when I called and had a good chat with
them in the dusk of the evening, when I no more
dreamt of her death, than of the breaking out of a
volcano in London.

Fanny and I had another heavy loss last year, in
the death of our dear, honest, good friend, George
Napier, a loss which we should have felt much more
if it had not happened so very nearly at the same
time with Mary's. Yet still, for all this, I must look
back on the past year as one in which I enjoyed

1874. much happiness, and had very much to be thankful
for.

Your visit to Rome with Leonard must have been
very interesting, and I daresay the sight of the
places, which we saw together in '48, brought back
many recollections. I am very glad to have seen
Rome, and seen it repeatedly and well before the
new people set about furbishing and smartening it
up. Parisianizing it, and improving away its
picturesqueness and peculiarity. They will not,
however, do away with the beauty of those views
across the Campagna to Tivoli or Albano.

I have been quite delighted with Mrs. Somerville's "Memoir;" it is long since I have read a
book that has pleased me so entirely. It brings
Mrs. Somerville, herself, before me, exactly as I
remember her. The editress (Miss Martha
Somerville, the daughter of Mrs. Somerville), too,
appears to me to have done her part remarkably
well.

John Stuart Mill's "Autobiography" made a
very different impression on me, though it likewise
is very interesting. It is the history of an extraordinary man and an extraordinary life, and written
apparently with great candour and honesty.—But I
read it with a continual mental protest against
almost all his ways of thinking. Almost all his
opinions are utterly antagonistic and repugnant—I
might almost say, detestable to me, and yet I do
not by any means think him a detestable character;
on the contrary, there is much in him that is
interesting and even attractive. He had a most

extraordinary education; it really almost makes 1874.
one's head ache, merely to read of the things he
was made to learn before he was twelve years
old. A great contrast to Mrs. Somerville, who
with the greatest difficulty got leave to study at
all.

While we were alone, I read aloud to Fanny the
first volume and the greatest part of the second of
Helps' "Spanish Conquest of America," which is
exceedingly interesting. I am also reading over
again Sismondi's " Républiques Italiennes.

Fanny's Journals keep you acquainted with our
doings, which since November have been very
quiet. We had delightful society in October—I
have seldom passed a more enjoyable summer.

Charles Lyell has wonderful energy; he is quite
devoted to his geology, which is now his chief source
of happiness; indeed, I think it mainly keeps him
alive. The clearness and strength of his intellect in
everything connected with that subject are still
remarkable. I am extremely glad that your
book on Florence has had so good a sale ; it is
indeed an excellent work and well deserving of
success.

With much love to dear Joanna and hearty
good wishes for you both,

I am, ever your affectionate brother,
CHARLES J. F. BUNBURY.

JOURNAL.

1874.   Lady Harriet Hervey and her daughter Isabel, the Barnardistons, Sir George and Lady Nugent, and two daughters, John Hervey arrived. Lady Charlotte Osborne, the Victor Paleys, and the Percy Smiths dined with us.

———

Fanny and all the party except me went to Mrs. Wilson's ball.

———

Lady Hoste came to luncheon. Mrs. Wilson, Lady William Graham, Lady Parker and Mrs. Montgomerie to afternoon tea.

Gave the Memoir of my Father to Lady Nugent.*

———

Very mild weather.

Walked with Lady Florence and Fanny.

Lady Cullum and Mrs. Dawson dined with us.

Most of our party went to the Bury ball—Lady Harriet, Sir George Nugent, Fanny and I stayed at home.

———

Our company all went away.

* Lady Nugent's father, Lord Colburne, was one of Sir Henry Bunbury's oldest friends.

The new Bishop of Ely, Doctor Woodford, came with Archdeacon and Mrs. Chapman to have luncheon with us.

It was the first time of our seeing him. His appearance and manner are in his favour, and we thought him very pleasant, and evidently a man of much cultivation. Mr. and Mrs. Palmer Morewood arrived

Clement went away.

January 19th.

Received an interesting though short letter from Charles Kingsley, who tells me that he and dear Rose expect to embark on the 29th, for America.

They intend to go as far as Colorado, if not as California, and hope to be in the west at the right season for seeing the "carpet of flowers" on the prairies. He concludes with an affectionate and touching farewell to me and my wife.

January 20th, Tuesday.

A very fine day.

Walked round the garden with Lady Mary, and many more of our party.

Lady Cullum and Headly Bevan dined with us.

I have not yet noted that Cecil and his very pretty bride Susie* (W. Napier's daughter), stayed with us from the 3rd to the 8th of this month. Susie is not only pretty, but intelligent, as well as sweet and gentle. From the 12th to the 16th,

* They were married in the spring of 1873.

1874. we had a large party in the house, for the occasion of two balls—Mrs. Wilson's and the Bury ball. The party consisted of Lady Harriet Hervey and her daughter Isabel, Sir George and Lady Nugent and two daughters, the Barnardistons, John and George Hervey, Alice Praed, Clement.

Wednesday, 21st.

A perfect spring day.

We planted a Tulip tree for the Morewoods, and chose places for some other trees. Had a pleasant walk with most of our party. The Hortons and Freda Broke, the Chapmans, the Victor Paleys, Mr. Lott and Mr. Hall dined with us.

Thursday, 22nd.

A very beautiful day, with white frost in morning. We planted trees for Lady Mary, the Miss Egertons, Minnie Powys, and Lord and Lady Rayleigh. The Morewoods and the Rayleighs went away after luncheon. Arthur went back to school.

Friday, January 23rd.

Another fine day.

The Egertons, Louisa MacMurdo and John Hervey went away.

We walked with Minnie Powys about the grounds. Read to Fanny and Minnie Powys in evening, Dean Stanley's Sermon at St. Petersburg.

Saturday, 24th.

Minne Powys went away.

News came of the startling announcement of the

immediate dissolution of Parliament. What a
bustle there will be all over the kingdom for the
next few weeks.

I read to Fanny Mr. Gladstone's address—or
manifesto — to the electors of Greenwich. Very
clever certainly. In the evening resumed the
reading to her of Helps' " Spanish Conquest."

27th, Tuesday.

Parliament dissolved—the elections to take place
before the middle of February.

LETTER.

Barton,
January 27th, '74.

My Dear Katharine,

I find by my register that I have not written
to you since December 24th, last year ; it can have
been only because I had nothing of particular
interest to tell you, as I trusted to Fanny to tell
you all about our company and *gaieties*. I am much
interested in what you tell us about Frank, and
I assure you I sympathise very much with you in all
your anxieties about him. I have no doubt it is
very difficult for a spirited, enterprising young man,
who is not fond of study, to find satisfactory employment at home in these days ; and besides, I do
not think it is strange or unnatural or blameable in
such a young man to wish to range and explore, to
see nature in distant countries, and to try some experience of a life less trammelled than ours.—So did
our English "gentlemen adventurers," in good

1874. Queen Bess's days.—So, though I feel for you and
your anxieties about him, I am not much surprised
nor at all shocked at hearing of the Natal scheme,
and I trust it will turn out very well. It is a great
comfort that there are friends of your own, and
such good people in that country, to whom he may
look for assistance and advice. The country is,
I believe, a very healthy one; and though I suppose
the chances of worldly success are uncertain in that
colony as in others, it seems on the whole to
promise as well as any scheme one could think
of. I do hope all will go well with him.

I was very glad indeed to have Leonard in our
party here; he made himself very agreeable, and I
was very glad to hear from himself of what he
had seen in Italy.

I have got Parlatore's Flora Italiana, as far as yet
published; it is likely to be voluminous, for these
five volumes (or rather four and a half), comprise
only the Monocotyledons and a small part of the
Dicotyledons,. It is evidently done with great
care, perhaps with excessive care, for the descriptions are minute and elaborate to a degree that
seems unnecessary; he is also a great multiplier both
of genera and species.

I am still studying Schimper's Paléontologie
Végétale, and think it very valuable.

The article on John Stuart Mill in the January
number of the *Edinburgh* appears to me most excellent, and there is in the same number an interesting account of the first Lord Minto. The
Quarterly on Mrs. Somerville is good.

Were you startled at the news of the sudden 1874.
dissolution of Parliament ?   We were; it was very
unexpected ;   it is *sharp practice*.   What a bustle
everybody will be in for some weeks to come !

What a wonderfully mild winter we have had
hitherto—thrushes singing and woodpigeons cooing
—hepaticas, snowdrops and violets in full flower.   I
do not know whether it is very healthy, or whether
we ought not to wish for Kingsley's favourite, the
"black north-easter."

Our party was a very pleasant one.   Lady Mary
Egerton is delightful, so are her daughters.

I hope you have been free lately from your enemy
the headache.

Believe me ever,

Your very affectionate brother,

CHARLES J. F. BUNBURY.

I shall like very much to hear how Leonard
gets on with his lectures.   I have no doubt he will
do them very well.

JOURNAL.

January 29th, Thursday.

Received a letter from Mr. Manning Prentice, of
Stowmarket, a leading political dissenter and active
Radical, inviting me to stand for West Suffolk,
on the *Liberal* (*i.e.* Radical) side.   Wrote to him,
declining and informing him that my political
opinions were quite different.

Friday, 30th.

Began reading Chief Justice's summing-up in Tichborne case.

Re-arranged some of the Compositæ in my herbarium according to Bentham.

Saturday, 31st.

A beautiful day. We were much out-of-doors.

Lord Augustus Hervey and Colonel Parker re-elected members for West Suffolk, without any opposition.

The weather during this month has been strangely mild, and in general very fine.—Not a flake of snow, and no frost sufficient to produce more than very thin and *short-lived* ice: on the other hand, very few really rainy days. Many days very fine, indeed spring-like and delightful.

The gardens are beginning to be quite gay with the early spring flowers, violets in profusion, snowdrops, the blue hepatica, anemones (A. coronaria), crocuses.

The birds are singing as if it were really spring, and I have repeatedly heard the ring-dove cooing, which I think is unusual in January.

February 1st, Sunday.

We received the Communion.

February, 2nd.

I am very sorry to find that Sir Bartle Frere, the highest authority, thinks the report of Dr. Living-

stone's death too strongly confirmed to be doubted. 1874.
So there is a most interesting and admirable man
gone to his rest. I am glad I have seen him,—
both at the Geological Club and at the British
Association meeting at Bath in 1864.

On the latter occasion I heard him deliver a
lecture in public, on Africa and the Slave-trade, and
both Fanny and I were much impressed by his
evident simplicity and earnestness.

Tuesday, 3rd.

A very dark, foggy day.

Wrote a note on a beautiful Odontoglossum
which has flowered.

February 4th.

The election at Bury, yesterday, ended in the
return of the two Conservative candidates. The
numbers were :—Mr. Greene, 1004 : Lord Francis
Hervey, 914 : Mr. Hardcastle, 707 : Mr. Lamport,
628.

This day I completed my 65th year, in good
bodily health, and I trust sincerely grateful to the
Almighty for the many and great blessings which I
am permitted to enjoy—for which indeed I can
never be sufficiently thankful.

This was a beautiful day, quite spring-like, and
we were much out-of-doors.

Went on reading to Fanny, Helps's " Spanish
Conquest of America."

We walked with Scott round the grounds, and decided on various operations, cutting underwood, etc.

A visit from Lady Cullum.

———

February 6th.

Fanny went to Mildenhall and returned to dinner.

———

LETTERS.

Barton,
February 7th, 74.

My Dearest Cissy,

I thank you most heartily for your truly kind and loving letter which I received yesterday, and which gratified me exceedingly.

I have a great many and very great blessings to be thankful for, but none greater than that of always finding so much kindness and affection in those who are nearest to me : and I hope I need not say that with all my heart I return your good wishes. I thank you very much also for the drawings,* which are an agreeable addition to my book. Fanny was much diverted by " Familiarity."

I wish you could have given a better account of Henry's health. I think the winter is generally a bad time for him (as it, in one way or another, is for most people who have not strong health), and one must hope that when *that* is over he will feel better, and that the air of Marchfield will do him good. I am not much surprised that he does not

* The drawings of his father.

take to Marchfield at first : I know it would take 1874.
me a long time to be reconciled to a new home.—
But it is a great comfort that you and Emily like it
so much, and find so many pleasant acquaintances
and such a good neighbourhood.

Fanny's letter will have kept you informed of our
very quiet goings-on here since our last party went
away ; we are very happy in our tranquility, and
never find the day long enough for our employ-
ments. She went yesterday to Mildenhall to visit
the new clergyman and his wife, Mr. and Mrs.
Robeson ; we are very much pleased with all we see
and hear of them, and they seem much pleased with
the place ; I hope that the choice (which caused me
much anxiety), will turn out very satisfactory, and
that Mildenhall will at last have the benefit of a
thoroughly good clergyman. Old Mr. Philips was a
terrible incubus upon it for a great many years.

I have not time to write more at present, except
that we are both well I am thankful to say.

With best love from both of us to Henry and
Emily, as well as yourself,

> Believe me ever,
> > Your very affectionate brother,
> > > CHARLES J. F. BUNBURY.

> > > Barton,
> > > February 8th, 1874.

My Dear Katharine,

I thank you very much for your kind letter
written on the 4th, and for your good wishes.

I do not at all wonder that your thoughts have

1874. been very much engaged by Frank, and the preparations for his departure, so as to leave you little power of thinking of anything else. He is, indeed, going a very long way off, and though it is to a very healthy country, and in good company, and to a country in which friends of your own are living, all of which are comforting circumstances, yet of course you cannot help feeling that it is a serious and anxious parting. But I trust you will see him return safe and well, satisfied with his wanderings, and improved in every respect.

Yes, it is very true, as you say, that dear Mary's death has impressed us very strongly with the feeling of the uncertainty of life : and yet, perhaps, the death of younger friends (of which we had two sad instances the year before), ought to impress that feeling still more strongly. But I am sure that the effect ought not to be to produce a morbid melancholy, or to interfere either with cheerfulness or industry.—Not that I have a right to say much on this last point.

I wish you could see our conservatory now : it is so beautiful, with Poinsettia, Euphorbia jacquinifolia, Eucharis, Begonia, Hyacinths, and a splendid crimson Tacsonia (a Passion flower, in reality), to say nothing of our old Cape friend, the *Pig-lily*. There is also a lovely white Orchid called Odontoglossum (?) flowering for the first time.

I will say no more at present, except that we are (I am thankful to say) both well, and that I am ever,

Your affectionate brother,

CHARLES J. F. BUNBURY.

JOURNAL.

Arranged some Compositæ according to Bentham.
The turn which the general Election has taken—
the success of the Conservatives—is very remarkable.

There can no longer be any doubt as to the
"Conservative reaction." Edward writes to me
from London, —"that the Metropolitan elections
have caused an immense sensation, especially those
for Westminster and the City, not merely on
account of the success of the Conservative candidates, but still more of the overwhelming ma-
jorities by which they were carried. When to these
are added Manchester and Leeds, at each of
which it is the principal man among the Liberals
that has been turned out, it is impossible to doubt
that there has been a great change of opinion in the
country, and a reaction against the extreme Liberal
party. The Conservatives now seem sure of a
majority of 30, if not more ; and with this number
they may very well look forward to at least three
or four years of office, if only they stick to their
colours as true Conservatives, and do not let
D'Israeli lead them astray with any of his Radical
devices."

He adds, that almost all his old friends are thrown
out:— Hardcastle, Headlam, Bonham Carter, Locke
King, who have been in the House since 1847. I

1874. am sorry, too, personally, for the defeat of a few of our Radical friends — Lord John Hervey, Sir G. J. Young, Hugh Adair ; — though not at all displeased at the general result. Even Gladstone only came in second at Greenwich, but I do not know that any of the Ministers have been actually thrown out, except Ayrton, whose defeat I cannot be sorry for.

LETTER.

Barton,
February 9th, 1874.

My Dear Leonora,

I thank you very much for your kind letter of the 3rd, and for your good wishes on my birthday. I have a great many causes for thankfulness, and among them, not the least, that Fanny and I both enjoy at present, very good health, and have had no occasion for medical advice (as yet) all this winter. After the long spell of mild weather it has turned suddenly very cold within these few days, and it looks as if winter were going to begin in earnest.

Our friends, Captain and Lady Florence Barnardiston, when they were here in January, were talking of going to Norway or Sweden to get some skating ! but it appears that even at St. Petersburg, the winter has hitherto been unusually mild ; while from Florence they write of remarkable cold. I am interested by your historical remarks. The struggle now going on in Germany between the secular and the papal power does strike one as having an 1874. analogy to that between the Emperors and the Popes of old. I feel just now rather familiar with the olden struggle, because I am reading Sismondi's "Républiques Italiennes," and have just come to the final destruction of the house of Suabia by Charles of Anjou, acting ostensibly on behalf of the Popes. I do not know whether the Germans consider Sismondi as a good authority. He is a very interesting writer, and appears to me honest and fair. For instance, while he warmly takes the part of the Italian Republics against Frederic Barbarossa, he points out that that Emperor though a haughty and ambitious despot, was not cruel according to the notions and habits of those times. To be sure, the standard of cruelty in the 12th and 13th and 14th centuries, was very high. I have not sufficient information to form a judgment upon what is now taking place in Germany; but I hope the Government wlll avoid unnecessary violence, and not lay themselves justly open to any accusation of tyranny or persecution.

I am reading to Fanny, in the evenings, the "Spanish Conquest of America," by Sir Arthur Helps, very well written, and extremely interesting —painfully interesting, It is quite melancholy to think of the miseries and destruction which fell upon those unfortunate natives of the New World; and yet, while one detests the cruelty and selfishness of the Spaniards, their courage and abilities were so striking and their achievements so extraordinary, that it is one of the most romantically

1874. interesting histories in the world. Prescott's "Philip the Second" is a very interesting book indeed: it is a thousand pities it remained so incomplete.

The strong turn which the General Election in this country has taken, in favour of the Conservatives, is very striking, and generally unexpected. Such a number of old members of the House who had sat in Parliament after Parliament for 20 years or more, have been thrown out—it is very surprising. I am sorry for several of them individually, especially for some of our own and of Edward's friends: but I am not at all sorry for the general result. I was not, however, zealous in the matter, because I feel little or no confidence in D'Israeli, and there is hardly a chance of seeing Lord Derby* (as I should wish) at the head of affairs. The change of Ministry in England, will, I should suppose, make no difference in our foreign relations.

I will leave Fanny to tell you the family news. With much love to Annie and Dora and kindest regards to Chevalier Pertz.

I am ever your affectionate brother,

CHARLES J. F. BUNBURY.

* The fifteenth Earl of Derby, who succeeded his father in 1869.

JOURNAL.

News of the death of Herman Merivale.

—— —

Examined a Begonia from the conservatory and determined the species.

—— ——

LETTER.

Barton,
February 12th, 1874.

My Dear Joanna,

    I thank you very much for your kind letter and good wishes on my birthday. I well remember that you were with us on that day last year, and that Fanny had a dangerous fall, and a narrow escape from hurting herself very badly indeed. I had not heard before of the " Memoirs of Madame Le Brun ; they must, by what you tell me in your letter, be interesting ; I remember seeing some of her pictures in the Louvre. I have lately got Professor Parlartore's " Flora Italiana," 4 vols. and one part of the 5th, which I believe are all yet published : it is evidently written with the utmost care and most elaborate detail, and will certainly be very useful for reference. I hope to make much use of it. When completed it is likely to be

1874. voluminous.   The first 3 volumes comprise the
whole of the Monocotyledons, which are very
interesting, and which have in general been rather
neglected by recent authors.

I was sorry to decline Professor Parlatore's
invitation, which was very courteous and flattering
to me; but the office of a *Juror* would not have
suited me at all, and we should not be able to go
abroad this spring.

I am reading to Fanny, in the evenings, Sir
Arthur Helps's " Spanish Conquest of America,"
a very good history, and deeply but painfully
interesting.   It is distressing indeed to read of the
cruelties and miseries suffered by the unfortunate
natives; yet the courage and ability shown by
the Spaniards was so great, and their adventures so
extraordinary, that scarcely any history has a more
romantic interest.   I think Cortes was as great as
a man can be without goodness.   Then the admirable—the really apostolic—goodness of Las Casas,
and some of the other clergy, affords a relief to one's
feelings; but again it is sad to see how their efforts,
and the good intentions even of the Government of
Spain, were baffled and defeated.

I have begun to read a second time, Mill on
" Liberty " which made a great impression on me
when I read it about ten years ago, and which
seemed to me to be unfairly treated in the
Edinburgh Review article on his Autobiography—
though in general I think that article excellent.

I am reading the concluding volume of the
"Life of Dickens," which contains some amusing

and interesting things, but also a vast deal of 1874.
verbiage.

(February 13*th).* The weather has changed again, and to-day is quite mild. Our outdoor flower beds were quite gay before the frost, with Snowdrops, Crocuses, Hepaticas and Primroses; and they seem likely to revive now that it has passed away. The conservatory is in great beauty. You will be surprised to see in the papers what a victory the Conservatives have gained in this general election ; it is a revival like that of 1845. It seems now clear they will have a good majority in the new Parliament.

I have not time to attempt a discussion of the causes of this great change of feeling. I am afraid *Beer* has had something to do with it ; but it is certainly by no means the only cause.

Believe me ever,

Your affectionate brother,

CHARLES J. F. BUNBURY.

JOURNAL.

February 15th.

News in yesterday's papers of a terrible fire in *our* quarter—the Belgravian quarter of London :— the Pantechnicon entirely destroyed, happily without loss of life, but with enormous destruction of valuable property, and in particular, I am afraid, of many valuable pictures which can never be replaced.

February 17th, Tuesday.

Fanny has had a letter from poor Lady Head,
speaking of the terrible misfortune that has befallen
her—the destruction of all her personal property in
the fire at the Pantechnicon. She had very lately
sold her house, and while looking out for another,
had stored all her goods for " safe keeping " in the
Pantechnicon, and all are destroyed. All the
collections of her own and Sir Edmund's life-time,—
books, pictures, drawings, letters, manuscripts,—all
the relics which no money can replace, and which
are so much more precious to one than money or
money's worth,—all are gone. It is most lament-
able.

———

Wednesday, 18th.

Visit from Lady Hoste.

Read Dean Church on Influence of Christianity
on Latin Races.

Gladstone has resigned, and D'Israeli has undertaken to form a new Ministry.

I am very glad that Louis Mallet has been
appointed permanent Under-Secretary of State for
India, in the place of Herman Merivale.

———

Thursday, 19th.

My Barton Rent Audit, and luncheon afterwards
with the tenants and Mr. Smith—all very satisfactory.

———

Saturday, 21st February.

Lady Susan Milbank came to luncheon with us, and Fanny went with her to call on Lady Cullum,— John Angerstein also came to luncheon.

—— ———

February 23rd, Monday.

The new Cabinet is completed, and the new men have kissed hands on their appointment, and the old set have kissed hands on resigning. The new Cabinet seems to be in general very good appointments, but it appears strange that the Home Office should be given to an unknown man like Mr. Cross. Perhaps no one more eminent was willing to take it.

—— ———

Thursday, 26th.

Furious wind and rain.

Mrs. Percy Smith sent us word yesterday of the death of Lady Northesk, the news of which had been telegraphed from Rome, and forwarded to *her* by her mother.* It appears to have been sudden. I am very sorry for it, though I have not seen Lady Northesk for more than 30 years. I have a most vivid and agreeable memory of her in the old Cape days, when she was Georgina Elliot, an intimate friend and constant associate of Sarah and Cecilia Napier. She was then remarkably beautiful, and altogether very attractive. For many years past, I believe, she has lived almost entirely abroad.

This morning I learn from the papers the death

* Lady Elizabeth Thackeray, a sister of Lord Northesk.

1874. of Lady Houghton*—but with her I was but slightly acquainted. Her death also appears to have been almost sudden. She was, I believe, an excellent person.

The last news from the seat of the Ashantee war are not entirely comfortable, though their first impression was cheering after the very alarming account which had come just before. Our army, it seems, had really gained possession of Coomassie, though after hard fighting, and with great loss of valuable lives, and the Ashantee king had declared his wish for peace, but how far he is really subdued, and whether we shall return to the coast without more fighting, seem doubtful questions.

It is certain that Sir Garnet Wolseley has shown very fine military qualities, and that the British troops have shown admirable courage and steadiness : but it seems clear also that the Ashantees are much braver and more formidable enemies than had been expected.

Up to London by St. Pancras, a comfortable journey : we found Dear Minnie Napier established at 48, Eaton Place. Later in the afternoon, Susan MacMurdo, Edward and William Napier came to see us. We learned from them the great news, that that enormous monster of a trial, the Tichborne

* She was Miss Crewe.

case, actually came to an end to-day:—that the
claimant had been found *guilty* on every count of
the indictment, and sentenced to 14 years penal
servitude. A most righteous verdict and sentence.

March 1st, Sunday.

Visit from Mr. Hutchings, Clement, Leonard
Lyell, Lady Arran and her daughter Mary.

March 2nd.

A visit from Charles Lyell, who seems in pretty
good health and spirits, though nearly blind, poor
fellow. He says, indeed, that he can still manage
to see the illustrations for his works, but he can see
nothing that is not quite close to his eyes, and this
inability gives an appearance of uncertainty and
unsteadiness to his walk, which inclines one to
suppose him much more feeble than he really is.

Charles Lyell has been much delighted by the
Duke of Argyll's Address to the Geological Society.
He was much pleased, some little time ago, by a
sort of address presented to him by the leading
members of the Geological Club, congratulating him
on having been 50 years a member of it.

One of the sayings now current is, in reference to
the General Election, that Beer has made the
Premier Dizzy.

Katharine and Rosamond came to luncheon.

Visit also from Montagu MacMurdo.

1874.   Lady Head and her daughter Amabel came to luncheon with us.

We had a little dinner party:—the MacMurdos and their pretty and clever daughter Mimi, Lady Head, Rosamond and Leonard Lyell, William Napier and Edward—all very pleasant. The MacMurdos and Edward in great force.

A fine day.

Henry came to luncheon with us.

We went to Burlington House, and saw the exhibition of Landseer's pictures. There is a surprising number of them, including sketches.

Many are very beautiful and delightful to look at, but on the whole I think it is rather a mistake to make an exhibition of the works of one painter only. There is an inevitable tendency to monotony, which is in danger of becoming wearisome. Speaking generally, I should say that *Landseer's* animals are admirable, and that his children, though inferior to the animals, are much superior to his men and women.

Fanny went to call on Mrs. Fazakerly.

Charles Lyell and his sisters drank tea with us. We dined with the William Napiers; met Lady

Bell, Miss Craig, the MacMurdos, Colonel Deedes, 1874.
and a Mr. Rudd.

Much talk about the last telegrams (just published) of the Ashantee war. The military men agreed that Sir Garnet Wolseley's last announcement, as to his halting on the return from Coomassie was very inexplicable, but they thought that he might be trusted, not to let himself be entrapped. Lady Bell is still a wonderfully handsome and agreeable old lady, of the age (I believe) of 86.

Sunday, March 8th.

A very fine day.

Read prayers with Fanny.

A visit from Louis Mallet, who (as I have already mentioned) has lately been appointed permanent Under-Secretary of State for India. He thinks very highly of Lord Northbrook, and has entire confidence in him ;—thinks it unfortunate that the English press have taken up the clamour against exportation of rice, in opposition to economical principles which were conclusively proved more than a generation ago.

He says the British Cabinet and the Indian Council were alike unanimous *against* prohibiting the export of rice.

The trade in rice by sea, from other countries, quite insignificant in comparison with the internal Indian trade.

If the rains were to fail a second season, it would be impossible for any Government to contend with the famine which would ensue.

1874.   I visited the Miss Richardsons, and had a pleasant talk with Helen,* who is a remarkably intelligent
and cultivated woman, and always very agreeable.

Mr. Reeve, the editor of the *Edinburgh*, being her
brother-in-law, she has the best opportunities of
hearing the news of the Literary world.   She told
me that the articles in the January number of the
Edinburgh, on "John Mill," and on the " Life of
Lord Minto," are by Mr. Reeve.

We have heard of a death under most sad and
distressing circumstances.   Mrs. William Nicholson
(sister in law of Mrs. Douglas Galton), has died
within a few days, from *blood-poisoning*, produced by
slightly wounding her foot with a small *copper* nail.

Monday, 9th March.

Much of the morning spent in giving directions
to the bookbinder.   Leopold and Lady Mary Powys,
the MacMurdos, William Napier and Clement
dined with us ; a very pleasant little party.

Powys remarked that nothing was said, in the
accounts from Africa, of wild beasts molesting the
march of our troops, or even observed by them ;
and that none even of our native auxilaries,
appeared to have suffered from snake-bites.   Mac-
Murdo thought that the country through which
our troops had marched, was too damp and marshy
to be suitable to venomous serpents.

He (MacMurdo) mentioned as a curious thing,
that, though he had spent several years in India, he

---

* Afterwards Lady Blackett.

never saw a Cobra de Capello, or any venomous 1874.
snake.

It very rarely happens that a European in India
is bitten by these creatures.

The great frequency of deaths from snake-bites
among the natives, is owing (he thinks) not only to
their going barefoot and bare legged, but also to the
usual situation and construction of their huts, which
are well adapted to be lurking places for such
reptiles.

March 10th.

Excellent news from Africa — we may hope that
the Ashantee war is virtually over.

Sir Garnet Wolseley's generalship, and the
soldiership of our troops seem to have been alike
admirable ; and though the war has been in itself a
miserable business, and any advantage we may have
gained by it a mere nothing in comparison to the
valuable lives lost, yet it is highly honourable to our
soldiers and sailors.

Still better news in the evening papers ;—all our
white troops safe across the Prah, on their return
march, and all ready for their embarking for
England.

Captain Glover's expedition successful as well as
Sir Garnet's—some of the native chiefs breaking
away from the Ashantee power. We may really
hope that this detestable war is now over.

March 11th.

The Louis Mallets, Lady Rayleigh (the Dowager) Richard Strutt, Edward, and Clement dined with us yesterday.

I asked L. Mallet his opinion about danger to our Indian power from a revival of religious zeal and energy in the Mahommedans; a danger which I have heard suggested by one or two persons.

He said, the Mahommedans of India had certainly been discontented, thinking themselves unfairly treated by our Government, — thinking that they were kept down and treated with disfavour in comparison with the Hindoos, whom they look upon as an inferior race. But nothing which had come to his knowledge led him to believe that there was any special religious excitement or agitation among the Mahommedans, or anything which threatened present danger. We dined with the Henry Lyells.

March 12th.

Minnie went to see the Royal procession; Fanny not being well, stayed at home; and so did I. The weather was dreadful,—furious snow-storms all the morning and mid-day; nevertheless, the proceession seems to have been a great success. Visits from Cissie and Emily, Charles Lyell, and his sisrer Marianne.

We all three dined with the MacMurdos at Fulham.

Saturday, 14th.

Fanny with the MacMurdos. Went to see Arthur.

Our dinner-party :—

Mrs. Ellice and her two daughters, Cecilia and Helen, General and Mrs. Rumley, Mr. Goodlake, Mr. Hutchings, Edward :—all very pleasant.

Mr. Goodlake is about to marry Cissy Ellice, to whom he has been engaged some years; he is not a young man, but is very lively, a great talker, and has seen a surprising variety of different and distant countries. He has been in not less than 30 of our colonies, besides many other countries whieh are not dependent on us.

Edward tells me of extraordinary and most interesting discoveries made by means of excavations on the site of Troy,—that is, on what has always been traditionally considered as on the site.

He is going to write an article on them in the *Edinburgh.*

Monday, 16th.

Mr. Walrond and the MacMurdos dined with us :—very agreeable. Mr. Walrond (as I think I have said before), is a special favourite of mine.

Curious accounts given by MacMurdo of the Duke of Cambridge's dislike to the Volunteer force in the early days of that "movement." If I understood him rightly, this feeling was grounded on political apprehensions, not merely on the military distrust of an irregular force.

MacMurdo still considers the Volunteer force as of great importance.

Mr. Walrond spoke favourably of Fitzjames

1874. Stephen's book on Liberty, Equality, and Fraternity;
and seemed to think it successful as a reply to Mill
on Liberty;—this latter he appeared to me rather to
undervalue.

Mr. Walrond spoke of John Mill's theory on the
"unearned increment of value" of land, as a wild
and most mischievous fallacy.

We agreed that a certain amount both of legal
and military training would be a valuable preparation for the career of a considerable land-owner.

At the Museum of Practical Geology, I saw a
satisfactory collection of specimens illustrating the
mode of occurrence of the diamond in South Africa.
The substance accompanying the diamonds in the
Vaal-River drifts (which are the richest) are small
pebbles (well rounded) of many varieties of chal-
cedony, agate, and jasper, and other forms of silica ;
none of the varieties of iron-oxyd with which it
is accompanied in Brazil. From the so-called "dry
diggings," there are specimens of the peculiar
decomposed greenstone, in which the diamonds are
there found, and the red garnets and green
diallage with which they are associated.

————

March 17th.

Drove out with Fanny.

Shopped, and visited Mrs. Edward Romilly, who
was very pleasant.

Edward dined with us.

————

Wednesday, 18th.

Visits from Miss Phillips, the Bishop of Bath and

Wells, Lady Head, Katharine, Cissy and her son 1874.
Willie, Mr. Hutchings, Mrs. Goodlake and Mrs.
Webbe.

Saw Charles Lyell at the Athenæum.

Thursday, 19th.

Emily Napier came to luncheon. I went to see
the South Kensington Museum.

Friday, 20th March.

That dear charming girl Minnie Powys came to
luncheon with us, looking very well and in great
beauty. Mr. Marcet also visited us. Charles
Lyell and his sister Marianne came in the afternoon.

The greatest part of these last 3 weeks the
weather has been mild, with very little of the
customary March winds, but there was one extraordinary burst of cold, beginning very suddenly in
the forenoon of the 9th, with snowstorms, which
went on at intervals through that and the following
day, with hard frost and piercing wind.

March 21st.

We returned to Barton—all well, thank God.

Sunday, 22nd March.

Read Kingsley's fine sermon on Death.
Fanny had a letter from Susan MacMurdo,

1874. describing the disasters which they have suffered (at Fulham) from the extraordinary high tide the day before yesterday.

Their garden was entirely flooded, and in fact destroyed—very deplorable, as they have been so hard at work restoring it and getting it into order ever since they returned from India. The water stood two feet deep in their kitchen.

———— ————

Monday, 23rd March.

A most beautiful spring day.   A long talk on business with Scott; signed the Churchwardens' Accounts.

———— ————

LETTER.

Barton,
March 24th, 1874.

My Dear Lyell,

I am sorry you have had trouble about the " Tables " of fossils in your new edition.   I had not looked at them before I received your letter ; because to say the truth, I am apt to find myself rather puzzled than enlightened by such-like " Tabular " formulas.   But immediately after reading your first note, I turned to Mr. Etheridge's *Tables* and examined them—and certainly I found several things to remark.

First — I suppose it is by a mistake of the printer, that *Monocotyledons* are treated as a subordinate division of *Acotyledons (i.e. of Cryptogamæ)*

a most extraordinary position. In these *Tables* 1874.
moreover they appear to be equivalent only to one
of the subdivisions of Dicotyledons, instead of the
whole : *Stigmariæ* is certainly not to be ranked as a
family, being only *roots* of plants; but this I infer
from your second letter, you have already discovered.

Cycadeæ and *Coniferæ* certainly ought to be
separated from *Monochlamydeous* Dicotyledons, and
put by themselves in a *gymnospermous* division.
You have already, in the body of the work pointed
out the importance of this group.

Anonaceæ are not Monochlamydeous, but are very
closely allied to *Magnoliaceæ*, and should come,
(according to the arrangement of the *Tables*) next
before *Nymphacaceæ*.

Nipaceæ are most certainly Monocotyledonous, and
appear to be intermediate between Palms and
Pandanæ.

Araceæ are marked in no formation earlier than
the oolite, yet at p. 424 you figure a fossil from the
coal-measures which is believed to belong to this
family.

Two families which, according to Heer, occur in
the Bovey Tracey Miocene beds, appeared to be
omitted in the *Tables*. They are : — 1. *Santalaceæ ;*
three sp. of *Nyssa* (*Tupelo* it is called in America).
2.—Rubiaceæ—one sp. of *Gardenia.*

My *Schimper* is unluckily in the bookbinder's
hands, so that I cannot refer to it ; else I might
perhaps have been able to give you fuller information ; but these are the principal points which have
occurred to me.

1874.   I do not see that you need correct or alter anything in consequence of what Professor Williamson
writes.   I am sure he will never be satisfied unless
you admit his infallibility, and this I see no sufficient
reason to do.   Believe me,

Ever yours affectionately,

CHARLES J. F. BUNBURY,

I received your first note only yesterday afternoon.

JOURNAL.

Thursday, 26th March.

Attended the parish vestry meeting; no remarkable business.

Fanny went to Mildenhall, and returned to
dinner.

Saturday, 28th.

Dear Minnie, Sarah and Albert Seymour arrived.

Monday, 30th.

Fanny with Minnie, Sarah and Albert went to
luncheon at Hardwick.   Susan MacMurdo arrived
—also the Barnardistons.   The Wilsons dined with
us.

Tuesday, 31st March.

A furious westerly gale in the night.
Wrote to Henry.

General Eyre came to luncheon. The Tom 1874.
Thornhills dined with us. Arthur MacMurdo
arrived.

LETTER.

Barton,
March 31st, 1874.

My Dear Henry,

　　　I have a bit of garden news for you : our
Aucubas are fruiting : two years ago, we bought from
Veitch two of the male plants lately introduced from
Japan and planted them close to two old bushes of
the female plant, one in the pleasure ground, the
other near the garden gate. Last year they flowered
and now the old females have a good quantity of
berries on them, fine, plump, well-grown berries, and
already turning red. I am pleased at the experiment succeeding so well and so speedily.

I was glad to hear General Eyre, who was here
just now, say, that he thought you looked remarkably improved in health.

Cissy's letters are so bright and cheerful, it is a
pleasure to read them. Susan MacMurdo, who is
here at present gives very animated and amusing
descriptions of the *flood* at Fulham (the extraordinary high tide on Friday the 20th), which was a
serious evil to them, ruining most part of their
garden, deluging their kitchen, filling their coal
cellar, and coming up almost to the drawing room ;
she says the children were delighted with the

1874. excitement.   It is a great pleasure to have dear
Minnie and the Albert Seymours with us now;
Sarah looking very pretty.

You have, I believe, heard of Charles Hoare's
dangerous illness, which has made, and makes us
very anxious, for dear, sweet Kate's sake.   We both
love her very much.   I am happy to say we have
had a postcard this afternoon, direct from Kate
herself, with a much better account : " he is going
" on very well, the fever has left him, and he is
" doing well in every respect."   This is a great
comfort.

I strongly recommend for your reading, " The
Life of the First Earl of Minto."   I have found it
an exceedingly agreeable book, both entertaining
and interesting.   Now, I am reading Sir George
Jackson's "Diary," (of the time from the Peace of
Amiens to the Battle of Talavera), and also a
volume of lectures by Mr. Maurice ; some of which
are admirable.

Our snowdrops and crocuses are now over, but
we have plenty of violets, anemones of two species,
primroses, hyacinths, forget-me-not and arabis ;
and the ribes sanguineum is coming into flower.
The conservatory is in great beauty.

Fanny is well, I am thankful to say ; so am I,
except an occasional touch of rheumatism.

With very much love to Cissy and Emmy,

Believe me ever,

Your affectionate brother,

CHARLES J. F. BUNBURY.

JOURNAL.

Lady Hoste and Lady Augusta Cadogan came to 1874. luncheon, and walked with us.

The Barnardistons went away.

———

Susan MacMurdo went away. Much talk with Scott on business, the Labourer's Union, etc.

A visit from Lady Cullum.

———

We went to morning Church. Took a walk with Minnie, Sarah, Albert and Louis Mallet.

———

We went in two carriages (Minnie and I in the brougham) to the Shrub, and admired the anemones and primroses.

A very agreeable party in the house this last week.

Dear Minnie, Sarah and Albert Seymour arrived on March 28th, are with us now, and will stay, I hope, all this week.

Sarah as charming as ever, and Albert very pleasant. The Barnardistons and Susan MacMurdo on the 30th, but *they* stayed only two days, and *she* only three, or rather two-and-a-half.

1874.   The Louis Mallets on the 2nd. Louis Mallet,
though a friend and admirer of John Bright, thinks
that the return of Bright into the Cabinet, was
among the chief causes of the defeat of the
Government in the late General Election—set the
whole of the clergy at once against them. He
thinks the influence of John Stuart Mill very
mischievous, and likely to continue for some time to
be so, especially as to political economy.

He quite agreed with me in thinking, that J. S.
Mill had made a great mistake in imagining that an
equal division of property could possibly be
reconciled with liberty ;—whereas (as he said and
as I say too), such a state of things can only exist
(if at all), under the most absolute and crushing
despotism.

He (L. M.) did not know J. S. Mill at all
intimately, but often met him at two political clubs
to which they both belonged. He says, that Mill
spoke extremely well at those meetings, with great
fluency and fine choice of language.

He remarked it as singular, that neither in Mill's
"Autobiography," nor in the "Memoir" of Grote,
is there the slightest mention of Anti-Corn-Law
League, or of that great struggle, or of the men who
were leaders in it.

Talking of India, he (L. M.), told me that the
population of the province of Bengal alone has now
been ascertained to be 64 millions. It had been
estimated at 42 millions, until last year, when a real
census was taken (a thing which had always been
declared to be impossible), and it was then found to

reach the number I have above given. The total 1874.
population of British India is probably not less
than 300 millions.

On every question relating to the condition and
government of India (L. M. says), he invariably
finds that men of the greatest ability and experience,
the highest authorities on the subject, are utterly
and irreconcilably at variance.

He thinks Lord Northbrook entirely in the right
in not forbidding the exportation of rice.

A good story about Mrs. Lowe (the wife of the
Ex-Chancellor of the Exchequer), told by Louis
Mallet :—

Mr. Lowe was expatiating on the absurdity of the
formula in our marriage service :—"With all my
wordly goods I thee endow."—For instance, he
said, turning to his wife, "when we married, I did
not endow you with all my my worldly goods, for I
had nothing." "Ah! but consider, my dear," (she
replied) "you had your great talents and learning."
"Oh!" said he, "I certainly did *not* endow you
with them!"

April 5th, Easter Sunday.

We went to morning Church and received the
Communion.

Monday, 6th.

A pleasant drive with Minnie in the brougham.

Captain and Mrs. Horton and the Victor Paleys
dined with us.

1874.   Tuesday, 7th.

Had a pleasant walk with Minnie and Sarah.

The Percy Smiths and young Milner Gibson dined with us.

Wednesday, 8th.

Scarlet fever at Haileybury. Arthur had to come home. Mr. Johnstone Bevan and his daughter Evelyn dined with us.

Thursday, 9th.

Albert Seymour was obliged to leave us, being recalled by the military authorities.

John Hervey arrived, and Mr. Abraham dined with us.

LETTER.

Barton,
April 9th, 1874.

My Dear Susan,

I have been excessively dilatory in acknowledging your kind and pleasant letter of the 4th of February: but I really am very much obliged to you for it, and it gave me great pleasure. I hope Joanna will now have quite got over her neuralgia, and that you are, both of you, well able to enjoy the spring, which is so beautiful in Italy.

You, I suppose, are just now very busy about the great Horticultural Exhibition at Florence, in which you are to take a part, and I hope it will all go off to your satisfaction. I have no doubt it will be a beautiful sight.

Our conservatory is in great beauty, and the out-door flower-beds are gay with hyacinths, dwarf-tulips, anemones, forget-me-nots, primroses and heartsease. We have had hitherto, not only a mild winter, but an uncommonly dry season, and an unusual proportion of fine weather, so that we are somewhat uneasy as to the supply of water for the summer.

Since we returned from London, we have con-stantly had company in the house, and therefore have not gone on with Helps's " Spanish Conquest of America," but have stopped (for a time), at the end of the 3rd volume. The peculiar characteristic of the book, that which distinguishes it from Robertson's and Prescott's and other histories of the time, is, that it is specially a history of the *natives*, of their treatment by the Spaniards, and of the efforts of good men to help and protect them. Las Casas and his brother monks play a much more conspicuous part in this than in any other history I have read, and a most noble one. It is comforting amidst so much that is melancholy and shocking, to read of the indefatigable labours of these good men. And it is striking to find the Dominican monks standing forward as the saviours of the Indians, when one has been used to associate this order with no other ideas than those of bigotry, cruelty, and the Inquisition. I confess I could not get on with Baron Stockmar's " Memoirs." I was much struck with the reviews

1874. of them, but when I tried to read the book itself, it appeared to me that the reviewers had picked out all that was interesting.

I have been much interested and entertained by the " Life of the first Lord Minto" (the father of the one who was so well known in Italy): it contains a great deal of curious information about the trial of Warren Hastings: the Regency question in 1788-9: the siege of Toulon in '93: the affairs of Corsica, &c., and the letters are written in a particularly clear, unaffected style. I find in it, too, several illustrations of Chancellor Oxenstiern's saying, by how little *(human)* wisdom the world is governed. Considering how the allied powers managed their affairs, it does not seem wonderful that the French Republicans were so successful. There are other illustrations of the same maxim in the Diaries of Sir George Jackson, which I have also been reading. He was a diplomatist, who was employed, first in Prussia during the disastrous time of Jena and Friedland, afterwards (under Mr. Frere) in Spain, during the early part of the Peninsular War. So much for books.

Fanny's journals will have told you of our visitors and our doings, (these last do not amount to much). It has been a great pleasure to have dear Minnie and Sarah staying with us for more than a fortnight past, and Albert Seymour* for the greater part of the time. Sarah is as pretty and as loveable as ever; still just the same sweet, light-hearted, affectionate, unpretending creature as we have always known her;

* Lord Albert Seymour, son of Lord Hertford, married Sarah Napier.

and we both of us like her husband very much. 1874.
Minnie, I am sorry to say, is by no means well.
Louis and Fanny Mallet, who spent about a week
with us, were very agreeable, and I may say, the
same of Lady Head, who is here now. It was a sad
misfortune, her having all her goods destroyed in the
Pantechnicon. I am so very sorry for her. Edward
had a narrow escape; he had had his whole collection of coins (which you know is very valuable) in
the Pantechnicon, and had taken it away just a
fortnight before the fire. By the way, Edward told
me, when we were in town, of very extraordinary
discoveries made by excavating on the site of
ancient Troy; and he has written an article on
them in the forthcoming number of the *Edinburgh*.
Katie Ambrose (MacMurdo) and her husband are
in England, and settled for the present at Southsea.
Leonard Lyell has gained his professorship, for
which he was eager, in the new University at
Aberystwith; I am very glad.

Give my love to Joanna, and pray remember me
to M. and Madame Parlatore.

Believe me ever,

Your affectionate brother,

CHARLES J. F. BUNBURY.

JOURNAL.

Saturday 11th April.

The 30th anniversary of our happy engagement.

Walked with Minnie and Sarah—we met Fanny
and Lady and Miss Head, and all took refuge in
the greenhouse, from a hail-storm.

1874.   Sunday, 12th April.

Did not go out—not quite well. Read Kingsley's fine sermon on "The Cedars of Lebanon," and Morris's "Lecture on Books."

LETTER.

Barton,
April 12th, 71.

My Dear Katharine,

I am extremely glad to hear that Leonard has been successful in gaining his professorship, for which he wished so much. I have no doubt that he is thoroughly well qualified for it, and will fulfil the duties exceedingly well. It is a good way to start in life, and may, I hope, serve as a step hereafter, to some more eminent position. Indeed, though the University is as yet new and little known, there is no reason why *he* may not render it famous. I have a high idea of what he is capable of accomplishing. I hope the place will agree with him, and that he will find people capable of appreciating him. Pray give him my hearty congratulations. I have not written to you since we returned from London, simply because I had nothing particular to say ;— I knew that Fanny's journals kept you informed of our visitors and our doings, and I had not been following up any special line of study. I have however been looking over and beginning to arrange my old notes on Botany.

I find it difficult to avoid having my attention distracted by the quantity of new books that are really worth reading. Our reading of Helps's "Spanish Conquests" has been interrupted for the last fortnight, as we have not been alone. It has been a great pleasure having Minnie and Sarah and her husband with us; and the Louis Mallets also were very agreeable, and now we have in their place, Lady Head and her daughter.

It has been a very fine spring hitherto, with a singular absence of east winds, and the flower beds in front of the house are very gay with beautiful flowers: hyacinths, dwarf tulips, anemones, primroses, daisies, iris, primula, and arabis.

I think I enjoy the beauty of spring, the flowers and birds, more and more almost year by year; and I am very thankful that we both of us hitherto have health to enjoy it.

Have you heard from Frank?—good accounts, I hope.

Ever your very affectionate brother,
CHARLES J. F. BUNBURY.

JOURNAL.

April 13th, Monday.

Very stormy, wet and cold.
Montague MacMurdo arrived.

Wednesday, 15th.

General and Mrs. Eyre came to luncheon and

1874. stayed most part of the afternoon with us. The Wilsons and Mr. and Mrs. Byron dined with us.

─── ───

Thursday, 16th April.

Lady Charlotte Osborne came to luncheon, and walked about the gardens with us.

───

Friday, 17th April.

We (Minnie, Sarah, Fanny and I, and Arthur) drove to the Shrub, and enjoyed the beauty of the wild flowers.

In the evening I read to the ladies, part of "As You Like It."

───

Saturday, 18th April.

Read some more of "As You Like It," in the evening.

Wilhelmina Legge, a really charming girl, is engaged to be married to Mr. Townshend Brooke.

I am very glad to learn by a letter which Fanny has this morning received from Lady Florence Barnardiston, that the match is one which gives great satisfaction to all the family—that Mr. Brooke is a country neighbour of Lord Dartmouth, and has long been known to them. We may therefore hope that he will prove worthy of her. It will be a loss to Octavia, for the two have for many years been inseparable.

═══

LETTER.

Barton,
April 18th, 74.

My Dear Lyell,

Many thanks for your corrected copy of 1874.
the *Tables* of Fossils which I have just received.
In looking through the tables of vegetable fossils,
I am much struck with the evidence of that
"imperfection of the Geological Record," which
you and Darwin have so often insisted on. The
gaps are striking. In particular, the Pliocene
column is almost a total blank as to vegetable fossils;
now one cannot believe that there was *no* vegetation
in the Pliocene period—that there was a great gap
between the Miocene flora and that of the recent
time, while there is so strong an analogy between
them. It can only mean that no plant bearing
fresh-water strata of the Pliocene age happens to be
preserved in our Islands. Indeed we should know
nothing (as far as Britain is concerned) of our
Miocene Flora, if those mere *scraps* at Bovey
Tracey and in the Isle of Mull had not happened
to be preserved. Of course I am quite aware that
what I am saying is not new, but it never struck
me half so forcibly before. In the animal world the
deficiencies are not nearly so striking.

What will Sir John Lubbock and Mr. Evans say
to Dr. Schliemann's discoveries on the site of Troy?
where he found an immense quantity of *stone*
implements at a small depth below the surface—and

1874. "*below* these, at a considerably greater depth, arms and implements of bronze in abundance, associated with elaborate works in gold and silver, as well as with finely-wrought pottery." I refer to p. 541 of the new (April) number of the *Edinburgh Review;* indeed the whole of that article ou Schliemann's Trojan antiquities is well worth reading. I believe it is written by Edward. I was exceedingly glad to hear of Leonard's success.

With kind remembrances to Miss Lyell, believe me

Ever yours affectionately,
CHARLES J. F. BUNBURY.

JOURNAL.

April 20th.

A beautiful day.

Spent the morning in writing a letter to *The Bury Free Press* for publication, on the distressing subject of " the Labour Question," *i.e.* the dispute between the agricultural labourers and their employers, which occupies so much of our thoughts and attention in this county.

I am strongly opposed to the lock-out system.

Tuesday, 21st April.

I dined with Maitland Wilson, by special and sudden invitation, to meet the Duke of Grafton, and talk over the *Lock-out* question. The party, a small one, and almost entirely masculine, Mrs.

Wilson and her sister, Mrs. Byron, the only ladies.
A long and amicable discussion, the Duke and I
quite agreed in opinion, and decided *not* to attend
the great meeting of farmers at Bury the next day.

Wilson, I think, was not far from agreeing with us
—but he thought it necessary to *humour* the farmers
more than appeared to the Duke and me to be
requisite.

April 22nd.

A beautiful day.

Fanny heard from Sarah Hervey.

We (Minnie and Sarah, Fanny and I), dined
with the Maitland Wilsons, and met the new Bishop
of Ely (Dr. Woodford) ; I had a good deal of conversation with him, and thought him remarkably
agreeable.

April 23rd.

Another very beautiful day, and very warm.

A pleasant drive in the open carriage, with
Minnie, Sarah and Fanny.

Lady Susan Milbank dined with us.

Friday, 24th.

Beautiful weather.

Dear Minnie and Sarah went away ; I need not
say that I was very sorry to part with them.   I need
not add much to what I have often said of them.
Sarah is as loveable as ever, not in the slightest

M

1874. degree spoilt or sophisticated; and Minnie is always
like the best and dearest of sisters to us.

Strolled with Fanny—saw the new foal.

_____

Saturday, April 25th.

My letter on the *Lock-out* question was published
in the *Bury Free Press.*

_____

Tuesday, 28th.

The Bishop of Ely came to us on Saturday, 25th,
accompanied by Archdeacon and Mrs. Chapman.
He and the Archdeacon were away most part of
Sunday, engaged in the Confirmation; but late in
the afternoon (the weather being glorious) they
strolled round the grounds with us for a considerable
time and we enjoyed much pleasant conversation
that evening also, as well as at breakfast both
mornings.

The Bishop is certainly an uncommonly agreeable
man—one of the best conversers I have met with;
a very winning manner, extensive reading and
variety of literary knowledge, abundance of anecdote
and a lively vein of humour.   Both in appearance
and in the style of his conversation, he strongly
suggested to me the idea of a fine old Benedictine
Abbot, French or Italian, of the 17th or 18th
century, one of those who combined learning with
geniality.—such as Sir James Stephen has described
in his essay.

Archdeacon Chapman also was very agreeable.

The Bishop and the Chapmans went away on 1874. Monday morning.

The Bishop praised Cardinal Wiseman's "Fabiola," as a beautiful little romance, founded on very early traditions concerning St. Agnes. He said that the apocryphal history of Paul and Thecla might be considered as one of the earliest of romances, as early at least as the second century, and long believed in the Christian Church. The edict of Diocletian (he said), which required the Christians to give up all their sacred books to be burned, first gave rise to a distinct examination of the books which had been in use among them, and a clear separation of the canonical from the apocryphal books; when they gave up the apocryphal to the Imperial officers, and carefully concealed the others.

The Bishop said that John Bright had, in conversation with one whom he, the Bishop knew, lamented his own want of a classical education; what he particularly felt as a deficiency was, the not having his memory sufficiently stored with reading to supply apt quotations.

The weather bright, but very cold.

Read chapter xi. of First of Kings, and chapter i. of Second of Kings, with Commentary.

A long talk between Fanny and John Phillips and me, about the labourers.

Sent order to Oakes and Bevan for payment of Vicarage rent to Mr. Robeson.

                     Wednesday, 29th.

Fine and bright but very cold weather.

Talk with G. Betts on Mildenhall business.

May 1st, Friday.

Very cold but bright.

We went early to Mildenhall, and spent the day there, returning to dinner. Board of Guardians—luncheon with Mr. and Mrs. Robeson, and much talk with them—they very pleasant. A pleasant letter from Henry.

May 2nd.

Cecilia Ellice was married on the 28th at Hampton Court Palace, to Mr. Edward Goodlake. Minnie and Sarah were at the wedding. It has been a long engagement, some years I believe, honourable to the constancy of both; and I trust they will now be happy. Cecy Ellice is a clever, agreeable, and interesting young woman. Minnie thinks highly of her, and I have much confidence in *her* judgment.

May 3rd.

Hail and rain—atmospheric phenomena which have been rare of late. This past month of April has been a very peculiar one as to weather; the drought quite remarkable. I think there were hardly more than three or four rainy days in the whole month, certainly not one in the last fortnight of it.

From the 18th to the 27th (both included) the 1874.
weather was steadily beautiful, quite like summer;
several of the days very hot. Since then, weather
bright and clear, with very cold wind; like March,
with a more powerful sun.

Read prayers with Fanny, and read to her
Kingsley's fine sermon on " The Mystery of the
Cross."

Tuesday, May 5th.

Weather a little milder.

Our good old Richard Palmer died. He had for
some time been in a declining state, but it was not
till lately that we became aware that his death was
near. He sank gradually, and died, I believe, of a
gradual failure of vital power, without any very
distinct malady. He was aged 69, I believe, but
looked older: he had been, I understand, in my
father's service since 1832, and continued as our
butler from 1860 till he was disabled by a paralytic
stroke in February, 1871 : since which he has lived
(with his wife) in one of the lodges of our park.—He
was an excellent man, earnest, simple-minded,
deeply religious, warm-hearted, with a strong sense
of duty, zealously and faithfully attached to our
family, and especially to his old master, my father,
on whose praises he expatiated to Fanny, one of the
last times that she visited him.

The last time that I saw him was on the 28th of
April, a week before his death, he was then in bed,
from which he never after rose : his mind appeared

1874. to be perfectly clear, but he was too feeble to talk much.

Inspected the books newly come from the bookbinder. — And the labels of trees for the Arboretum.

The Maitland Wilsons (including Agnes), Mr. and Mrs. Byron and Lord John Hervey dined with us. It was pleasant.

I find I have omitted to notice, in its place, the sudden and startling death of the famous geologist, John Phillips, last month. He was killed by falling down stairs on leaving a friend's room at Oxford. I was very sorry :—he was a geologist of a high order, and connected with the old *heroes* of geology, and especially interesting as the nephew and pupil of William Smith ; and he was, moreover, a remarkably pleasant and genial man, not solely a geologist, but intelligent and active-minded in various ways. I saw most of him in 1850, when we visited him at York, in our way to and from Scarborough : he was very obliging and attentive to us, and I learned much from him as to the geology of Yorkshire.

Up to London.

Had the delight of seeing Sarah Hervey and Kate Hoare. Sarah came at 2, with her brother George, and had luncheon with us, and Kate came in afterwards. It is always delightful to me to be in their company.

Visited Mr. Twisleton and Norah Aberdare —both very pleasant.

Charles and Marianne Lyell came to see us.

Thursday, 21st May.

Edward dined with us.

Katharine Lyell tells me that she has heard from her friend Lady Smith, who, some days ago completed her 101st year, and is still in full possession of her faculties. We heard also to-day from another visitor (Mrs. Johnstone), that Canon Beadon, of Wells, is in excellent health and spirits at the age of 97.

Friday, 22nd May

Visited Lord Hanmer and Lady Mary Egerton.

We went to the British Museum and saw the bronzes and gold ornaments of the Castellani collection—very interesting. The gold ornaments, necklaces, etc., are of most elaborate and exquisite workmanship, and in vast variety. The bronzes wonderfully fine :—I do not know that I have seen any like them except in the Naples Museum. Those I particularly remarked were:—A head of a goddess (Aphrodite?) of extraordinary beauty and

1874. grandeur—in a grander style, it strikes me, than
Aphrodite is commonly represented, with a something of melancholy in the expression. A *strigil* of
bronze (for use in the bath), with a very beautiful
naked figure of Aphrodite for its handle—she is
represented holding a similar strigil. A beautiful
figure called Orestes, taking refuge at the altar—I
should have called it a warrior falling.

Mr. Poole, who shewed the collections to us, is
most civil to us.

_____

Saturday, 23rd May

Cold, raw, and disagreeable.

Went with Fanny to the Zoological, and saw Mr.
Bartlett, the Superintendent,—then to Katharine's
and had a pleasant talk with her and her husband,
Rosamond and Arthur and Miss Stirling.

Minnie and her brother John dined with us, very
pleasant.

_____

Sunday, 24th.

Read to Fanny Kingsley's beautiful sermon on
The Spirit of Whitsuntide, from his new volume
of Westminster Sermons.

Visits from William Napier and Charles Lyell
and his sister Marianne.

Called on Lord Waveney (Shafto Adair), found
him confined to the sofa by an attack of gout, but
very lively and animated, and as usual very ready to
talk. We talked much about the Lock-outs and
Strikes in Suffolk, and agreed very well.

We dined with the Arthur Herveys, at 37, Dover Street (the Bishop of Ely's house, which he has let to them for a time),—met the Palmer-Morewoods who are staying with them, the Locke Kings, Lady Clancarty, Mr. and Mrs. Austen Leigh, Mr. and Mrs. Singleton, Miss Rodney, Lady Adeliza Hervey.

Mrs. Austen Leigh is daughter of the Archbishop of Dublin (Trench).

Tuesday, 26th May.

Cissy and Emmy came to luncheon—also Guy Campbell.

Visits from Mrs. Ford and Mrs. Coltman.

Went with Fanny to Dickinson's (about Mary Lyell's picture). Minnie and Clement dined with us.

May 27th.

My nephew Henry had brought a *monkey* from Malta, as a present to Fanny!—but as it was inconvenient to keep in the house, we drove off with it to the Zoological gardens, and saw it safely deposited there under the care of the keeper of the monkey-house.

It is a very young animal, of the species Bonnet Monkey, Macacus . . . . and is at present a most gentle, tame, confiding little creature, not showing yet any symptoms of mischievous or vicious propensities.

Our dinner-party:—Sarah and Albert Seymour,

1874. Sarah Hervey and her youngest sister, and their brother Sydenham, Minnie Napier, May Egerton, Mr. Sidney Osborne (son of Lord Sidney), Guy Campbell, and our nephew Henry.

An exceedingly pleasant party. I sat between the two Sarahs, both most charming women. Cissy and Emily, and the Palmer-Morewoods came in the evening.

———

Thursday, 28th May.

Henry came to see us.

I called on the Miss Moores and Mr. Marcet, and had pleasant talk.

Our second dinner party :—Lady Mary Egerton and her daughter Charlotte, Lady Head, the MacMurdos, Mr. Twisleton, George Hervey, and young Henry. Very pleasant.

Mr. Twisleton is a great talker, and a very entertaining one, full of anecdote.

———

Friday, 29th.

Sarah Hervey, two Miss Egertons, and Cissy came to luncheon with us. After luncheon, we two with Sarah drove out, saw Mr. Theed and his studio, and paid a long and very agreeable visit to Lady Lilford (the Dowager), and dear Minnie Powys, who is always charming.

———

May, 30th.

The 30th anniversary of our happy marriage.

Thanks be to God, who has allowed me so many
years of wedded happiness, so long a period of
harmonious union with so excellent and so admirable
a wife.

It was a beautiful day, and we drove out with
Lady Head and her daughter Amabel to Fulham,
and spent the afternoon very agreeably with the
MacMurdos, who were as usual very cordial and
very pleasant.

Their house (Rose Bank), is situated close to the
edge of the river, and hardly above its level, in
a situation which, in this fine summer weather, looks
very pleasant. It commands a good view across,
and along a fine reach of the Thames, which at the
time of our visit was "without o'erflowing, full"—its
surface glittering in the sunshine and enlivened
by many boats; its banks richly fringed with trees.

The MacMurdo girls, 5 in number, are all pretty;
the eldest, Mimi; very clever and interesting, with
a remarkable talent for drawing. (I think I have
mentioned her before). Susan, the second, uncommonly pretty, with most beautiful eyes.

———

May 31st.

Read a very noble and beautiful sermon of Kingsley's, on Prayer, one of his Westminster sermons.
Visited Mrs. Douglas Galton and Mrs. Berners,
very pleasant. Mrs. Hugh Berners, whom I visited,
showed me an old silver toilet service, to which a
curious story attached — how it had long lain forgotten, unowned, and unknown in the Bank of Eng-

1874. land ;—so long that the wooden box in which it was
enclosed fell to pieces on being moved; how the
ownership was discovered by means of some old
love-letters which were shut up in one of the caskets
—letters from the Mr. Berners of that day to his
betrothed ; and how by the production of wills and
deeds and family papers, the present Mr. Berners
established his right.

This silver toilet set is very handsome, and
belongs, if I understand rightly), to the time of
Charles the Second, or a little later.   Mrs. Berners
said that the love-letters were much more ardent
than those of the present day.

————

June 1, Monday.

Went out with Fanny, called on Mrs. Locke King
and on the Herveys, and had a very pleasant chat
with Sarah.

Henry Lyell and Katharine, Rosamond and Miss
Stirling, Cissy and Emily and Edward dined with
us.

————

June 2nd.

A very warm, bright day.   Fanny went with
Charles Lyell to the Woking cemetery and returned
to dinner.

Minnie came to luncheon with me, and afterwards
took me to the Athenæum.

————

June 3rd, Wednesday.

We walked round the Albert Monument in Hyde
Park, looked with some care at the sculpture, and

admired it much.　Of the four large groups at the 1874.
corners, the two we liked best were "Asia" by
Foley, and "Africa" by *Theed*.

The figures of artists and philosophers forming a
series, are fine, characteristic, and well-grouped.

———

Thursday, 4th.

Looked into the Royal Academy Exhibition,
but could not get a satisfactory look at *Miss
Thompson's* picture ("The Roll-Call after a battle in
the Crimea,") because of the crowd before it, though
there was no crowd elsewhere in the rooms.　I was
charmed by a picture of a girl by *Millais*—a lovely
picture of a lovely girl, called "The Picture of
Health."

Admired also two beautiful girls by *Frith*—
"Pamela," and "Wandering Thoughts;" a most
beautiful landscape, truly and characteristically
English, by *Vicat Cole*, "The Heart of Surrey:" a
grand wild scene of cliffs and breaking waves and
sea-birds on our northern shores, called "Our
Northern Walls," by *P. Graham : Millais's* "North
West Passage," a fine, rugged, weather-beaten,
earnest old sailor listening to something which his
pretty daughter or granddaughter is reading to him
(evidently something relating to the N. W. Passage)
and which excites him to a sort of grim enthusiasm.
—(A striking and interesting picture).

———

Friday, 5th June.

A splendid day, very hot.

We passed a delightful afternoon at the Zoological

1874. Gardens, with a party composed of Minnie Powys, Lord and Lady Arthur Hervey, Sarah, Patience Morewood, Caroline Hervey, Mr. Hutchings and Sir William Medlycott. Minnie Powys went with us from this house, we joined the Herveys in Dover Street, and the two gentlemen came up with us in the Gardens. There is always a great deal here that is worth seeing. We saw most of the beautiful or curious and amusing birds and mammals which we had seen in former visits last year and the year before:—the pelicans, flamingos, guillemots, hornbills, sun-bitterns and others. But the poor chimpanzees are dead.

We saw also a young lion, with which Sarah Hervey had made acquaintance last year, at the house of the Dowager, Lady Winchelsea, whose son, Mr. Finch Hatton, had brought it from Africa. It has grown very fast since then.

Interesting objects to look at, beautiful weather, and above all delightful company, made this a memorable afternoon for me.

Sarah Hervey and Minnie Powys both are as charming as it is possible to be.

A very curious bird which was new to me, was the *brat-bill*, apparently one of the Waders, with the head and beak so very oddly and grotesquely shaped, that it looks like a caricature of a bird.—It is the *Cancroma cochlearia*, Linn., nearly allied to the herons and bitterns, as I find by the Zoological Society Catalogue :—allied also to that still more strange creature the *Balaeniceps* which I saw here some years ago.

Miss Meade, Admiral Spencer and Clement came to luncheon.   Arthur arrived—we went with him to call on Lady Smith, then to see Henry and Emily. Fanny went out to a *drum* at Lord Hertford's—I stayed at home.

———

June 8th.

Our dinner party:—the Bishop of Bath and Wells, Lady Arthur and Sarah, Minnie Powys, Lord and Lady Hanmer, the Frederick Greys, Sir Bartle and Lady Frere, the Lynedoch Gardiners, Minnie Napier, Edward and our nephew Harry, all pleasant. Afterwards we had a *drum*, as the phrase now is—a large evening party, about 200 had been invited, and about 100 came.   This *drum* is an old fashionable phrase now revived, I never heard it in my youth, but one meets with it very often in the writings of last century.

The Bishop only came in for a time, from the House of Lords, where the important debate on the Public Worship Bill was going on, he came in after the rest of us had begun dinner, and went away before we had finished.

———

June 9th.

The Frederick Greys breakfasted with us, and were very agreeable.   Talking of the Archbishop's Bill, for regulating public worship, Sir Frederick said that most laymen whom one meets with seem

1874. to approve of it, yet he understands that the Dissenters mean to oppose it in the House of Commons; probably, as he said, because they think it will tend to strengthen the Church. In like manner, the Free Kirk people in Scotland oppose the abolition of Patronage in the Established Kirk; and Sir Frederick Grey told us that Lord Grey said on this subject to Lord Dalhousie . (who is one of the champions of the Free Kirk)—" the fact is, you are afraid of the opposition shop becoming too popular."

A pleasant afternoon party at the Arthur Herveys —a sort of leave-taking of their friends in London.

We dined with Lady Mary Egerton, Sarah Hervey going with us; met Miss Stanley, Mr. and Mrs. Spottiswoode, Mr. and Mrs. Dundas, a young Mr. Emmons, from America, who is acquainted with the Kingsleys.

The Egerton girls are very pleasant, intelligent, and highly cultivated.

_____

June 10th.

We went to have a farewell glimpse of the Arthur Herveys, who, I am very sorry to say, were to leave town the next day. It has been a great delight to see something of them, and especially of Sarah, in these last three weeks.

Then to Madame Christine Nilsson's concert at the St. James' Hall—admired some of her singing very much. She is decidedly handsome too—as Fanny said, a northern peasant beauty.

We went with Minnie Napier and Minnie Powys 1874. to Sir Bartle Frere's geographical evening party or reception, at Willis's rooms. A large assemblage ; we met many friends and acquaintances, and there were a good many strange and peculiar-looking people ; many curiosities from various parts of the world, and a fine display of large and instructive unpublished maps, especially some of the Lake region of Central Africa.

Some of these maps showed clearly the contrasted opinions of Sir S. Baker and of other African travellers — the former making, Lake Tanganyika communicate by a direct channel with the Nyanza, —the others representing a chain of hills between them.

———

Thursday, 11th June.

The MacMurdo's and Cissy and Emily came to luncheon.

We dined with the Bishop of Winchester (Dr. Harold Browne) in St. James' Square. The house is a very fine one ; the rooms in a grander style than one often sees in London houses.

We met Lord Onslow and his mother Mrs. Onslow, the Dean of St. Paul's* Mrs. Byron, etc.

———

Friday, 12th.

Sent Maurice's " Friendship of Books " to Sarah Hervey.

Louisa Hervey† came to luncheon.

* Church.　† Afterwards Mrs. Cyril Graham.

N

1874.   We went with Minnie to Lady Burdett Coutts
garden party at Holly Lodge. On our way back
called on Katharine.

———

June 13th.

Arthur went away.

A visit from Mr. Lombe.

We, with the two Minnies, went to Russell Square,
and saw Waterer's rhododendron show — very
beautiful. Lady Mary and Miss Egerton joined us
there, and brought me back.

———

Sunday, June 14th.

We went to St. Mark's Church, Notting Hill, to
hear Mr. Archer Gurney, who preached a fine
and vigorous sermon. As it was *Hospital* Sunday
there was a collection after the sermon. The service
was choral, with a great deal of music, choristers in
white, altogether like cathedral service; but in the
sermon there was nothing especially high-church,
indeed I believe Mr. Archer Gurney's doctrines are
a mixture of the high and the broad.

Notting Hill was to us, quite a new quarter, and
we were much surprised to see its great extent, and
the long ranges of handsome and well-built houses,
evidently fitted for the residence of wealthy people.

———

June 15th, Monday.

Received a charming letter from Sarah Hervey.

Mr. Walrond dined with us—very pleasant—my 1874.
nephew Harry also.

_____

June 16th.

Excessively cold.

We went, with Minnie Powys and Minnie Napier
to Holland House, Lady Lilford having got leave
from Lady Holland for us to see it. Mr. Lane,
Lady Holland's agent, a very intelligent and well-
mannered person, showed us through the whole of it,
and we saw it with every advantage of quiet and
comfort. But I cannot at all undertake to describe
it, in fact, though delighted, I was rather bewildered
by the number of rooms we saw, the intricacy of
their connections, and the splendour of their
contents. I will only say, in general, that it is a
beautiful house, and prodigiously rich in objects of
interest. I was especially attracted by the long
library, which Macaulay has commemorated in such
a striking passage of his Essays.

The room in which Addison died has also a
special interest. We were delighted also by the
sight of his writing table.

Of the pictures, some that I particularly remember
are—*Reynolds's* famous " Baretti"; a portrait of
Panizzi, by *Watts*, painted at the instigation of the
last Lord Holland, avowedly in rivalry with
Reynolds's "Baretti, and certainly very clever ; the
group (well known to me from the engraving) of
Lady Sarah Lennox, Lady Susan Fox Strangeways,
and Charles James Fox,* also what is very interest-

* Lady Sarah in this picture is much less beautiful than in the full length at
Barton, and still less so in the original sketch.

1874. ing, *Reynolds's* original small sketch for this picture;
several portraits of Charles James Fox, at different
ages; Caroline, the first Lady Holland, by *Reynolds*,
a beautiful picture; the same Lady Holland, at a
more advanced age, by *Ramsay*, (we have a copy of
this); Stephen, second Lord Holland, (I do not
remember to have ever before seen a portrait of
him); Elizabeth Lady (Vassal) Holland, (quite
lovely); a curious picture by *Hogarth*, of a performance of private theatricals.

We dined with the Baring Youngs — met Sir
George Lawrence, Lord and Lady Sidmouth, and
others.

Wednesday, 17th June.

Went shopping with Fanny, chiefly to Jewellers'
shops.

Minnie came to stay in the house. —Clement
dined with us.

Thursday. 18th.

Cissy and Emily came to luncheon.

We dined with the Locke Kings—met Lord
Devon, Sir Fowell Buxton, Sir Robert and Lady
Collier, Mr. W. Hoare, and others. Lady Collier
told me a good saying which is ascribed to the
Queen. When she was told that Sir Charles Dilke
had spoken disrespectfully and disloyally, she is
reported to have said " Poor fellow! I remember I
stroked his head when he was a 'little boy; I am
afraid I must have stroked it the wrong way."

Lady Collier is a clever, lively, entertaining 1874. woman. One expression she used, struck me as very happy.

Speaking of the difficulties of the French in settling their Government, she said:—" Why can they not have that which *we* find answer so well —a Republic with the forms of a monarchy ?"

Friday, 19th June.

We had a very pleasant little luncheon party:— Octavia and Wilhelmina Legge, Sarah and Albert Seymour, Minnie Powys, my niece Emily, Isabel Hervey and her brother Dudley, Admiral Spencer, Sir John Shelley, Richard Strutt, Cecil, Clement, and Harry.

Again to the Royal Academy, admired *Millais's* " Scotch Firs," and still more his "Winter Fuel," —trunks and branches of birch trees, heaped on a sledge to be brought down for fuel, with a bold, heathy hill for background — thoroughly true to nature, and highly finished in detail, with a beautiful general effect ; also *Hook's* " Jetsam and Flotsam," one of his characteristic Cornish or N. Devon sea-coast views ; " The Pass of the Cateran." a fine Highland scene by *Smart*. *Watt's* portrait of John Stuart Mill is impressive—strikingly conveying the idea of intense thought and feeling, with a settled look of melancholy and dissatisfaction.

Saturday, 20th June.

Dear Minnie Napier went off on her way to Carlsbad.

We dined with the Maitland Wilsons, — met Lord and Lady Gwydir, Lord and Lady Middleton, Lord Crewe, Lady Hoste and Mr. Greene, Herbert Praed.

———

June 21st, Sunday.

We went to the Savoy Chapel to hear Mr. White, who gave us an admirable sermon. His text was from the xiv. St. Mark—the passage relating to those who blamed the woman who annointed the head of our Saviour with the precious ointment.

The sermon was directed against the hard, cold, carping, censorious or niggardly spirit, which blames a generous expenditure for patriotic or religious objects which objects to monarchy as *too costly* and the like. That part of the sermon in which he extolled patriotism, and defended the institution of Monarchy, was especially excellent. Altogether, it was bold, animated, varied, and unconventional : in parts truly eloquent, while some passages would have been blamed by certain people as too much bordering on the humorous ; but I like this variety and boldness in a sermon.

———

Monday, 22nd June.

A fine, warm day.

Went to the Linnean Society's new rooms, at Burlington House, and brought away the volumes

of the *Transactions* due to me,—afterwards to Wynd- 1874.
ham Place to see Henry and Cissy, and to say
farewell to Harry.

––––––

June 22nd.

My nephew Harry left London to rejoin his
ship, the *Lord Warden*. He is an exceedingly
amiable and interesting lad, sweet-tempered, gentle-
mannered, warm-hearted and very intelligent besides.

Edward and Clement dined with us.

––––––

Tuesday, 23rd.

Two dear and charming friends came to luncheon
with us—Kate Hoare and Caroline Napier, Kate
looks blooming, and seems in good spirits and
happy about her husband, though his health is not
yet so completely restored as to allow of his
returning to business. Caroline Napier is, as she
always has been, most sweet and charming in
manner, and admirable in disposition and character,
and she is still very pretty, though her age cannot
be far short of fifty.

Our dinner party :—the Aberdares, Mr. and Mrs.
Drummond, General and Mrs. Eyre, Mr. and Mrs.
Angerstein, Mr. and Mrs. Longley, Lady Head,
Lady Rich, Isabel Hervey and her brother Dudley,
Mr. Theed and Edward—very agreeable. I sat by
my cousin Norah (Lady Aberdare), who is extremely
agreeable as well as excellent, and has a very fine
intellect.

June 24th, Wednesday.

We went to Mrs. Forde's afternoon party—
crowded rooms—profusion of works of Art : afterwards left cards at various houses, and ended with
half-an-hour in the Royal Academy.

Thursday. 25th.

Lady Augusta Cadogan, the MacMurdos and
Lady Hoste came to luncheon. We went to Mrs.
Tait's afternoon party at Lambeth Palace. Miss
Phillips, Miss Kinloch, Isabel Hervey and Clement
dined with us.

Friday, 26th June.

We dined with Lord Hanmer.

Fanny went with Minnie Powys to the Aberdares'
ball, and Isabel Hervey went to Lady Braybroke's.

June 27th.

Louisa Hervey, her betrothed Mr. Cyril Graham
and her brother Dudley came to luncheon with us :
Isabel Hervey was staying with us.—a very pleasant
little party. We thought Mr. Cyril Graham very
agreeable : he has been a great traveller in the
East, has resided much in those countries, and
seems to have a great knowledge of Oriental
languages. What is not always the case with great
Orientalists, his manner is particularly pleasing and
gentle. We had very good talk, though of course,
as to Oriental matters, I could only learn.

We drove out with Isabel Hervey, called on the 1874.
Legges, and saw them all, and also Wilhelmina's
wedding presents, afterwards Fanny and I called on
Mrs. Maurice.

Sunday, 28th June.

We went to Quebec Street Chapel, and heard a
very excellent sermon from Mr. Holland, on
domestic affections and duties, and on the religious
obligation to be kind and courteous at home and in
one's family.—Afterwards to see (the Dowager)
Lady Lilford, with whom and with dear Minnie
Powys we spent a good part of the afternoon very
pleasantly. Lady Lilford is full of cordiality and
warmth of feeling towards us. She is a true Fox in
warmth of heart.

June 29th.

Our dinner party:—Lord and Lady Rayleigh,
Leopold and Lady Mary Powys, Lady Mary
Egerton, Sarah and Albert Seymour, the Mac
Murdos, the Douglas Galtons, Mr. and Mrs.
Clements Markham, Mr. W. Gurdon, Leonard
Lyell.

Tuesday, 30th.

Fanny went to see Leonora and her children.

We went out with Cissy and called on Lady
Dalrymple Hay: afterwards we went, by invitation
from Lady Augusta Stanley, to the garden of the
Deanery at Westminster, to see an exhibition of

1874. Window-gardening by the poor.   It was a pleasant sight, and we met many acquaintances.

———

Mrs. Liddell, Mrs. Mills, Minnie Powys and Sarah Seymour came to luncheon.   We went to the Hookers' garden party at Kew, and much enjoyed a stroll in those glorious gardens.   The afternoon was beautiful (there had been a considerable fall of rain in the morning and the previous night), neither too hot nor too cold, bright and clear, with a pleasant breeze, and varying lights and shades from the passing clouds.   We were received at first in a small room on the ground floor of Hooker's house, but soon passed out from the back of it into the great gardens which were in full beauty and most enjoyable.   We had not time, however, to look into any of the plant-houses.

At the Hookers', had some conversation with General Strachey, a noted Indian explorer, a geologist and a very able man.

I learned that the Professorship of Geology at Oxford has been given to Prestwich, a very proper appointment.

We dined with Lady Rich : met the Douglas Galtons, Mr. and Mrs. Roundell, Mr. Heathcote, Mr. Robarts and a few more—pleasant.

———

A beautiful day.

Leopold and Mary Powys came to luncheon— very pleasant.

Visit from Mr. and Mrs. Robeson.

Drove out with Fanny: we picked up Isabel Hervey* and paid several visits, saw Lady Cullum and Mrs. Drummond and Mrs. Kay.

Cissy and Emily and William Napier dined with us.

———

Friday, 3rd July.

Saw the Water-colour Exhibition.

———

July 4th.

Leonard Lyell was married to Mary Stirling:— the ceremony performed by Mr. Latimer Neville and Mr. Symonds, in St. Michael's Church, Chester Square; the wedding breakfast at Lady Head's, where, also, the wedding presents were exhibited.

Lady Head has been a most kind friend to Miss Stirling and her mother, having known the family originally in Canada.

I wish all happiness to Leonard and his bride, and there seems good reason to hope it.

———

Monday, 6th July.

Settled accounts with Mr. H. Lofts for my London houses.

Our dinner party:—The Maitland Wilsons, the Gurdons, the MacMurdos, Mr. and Mrs. Vernon Harcourt (*she* was Rachael Bruce), Admiral

———

* Now Mrs. Cyril Locke.

1874. Spencer, Mr. Goodlake, Susan and Isabel Hervey,
Helen Ellice, Mr. Hutchings and Major Grant.—A
pleasant party.   In the evening came in some other
pleasant people, in particular, the Bishop of Bath
and Wells, the Hookers and Minnie Powys.

————

Isabel Hervey (Lord Charles' youngest daughter),
who has been staying with us since the 20th of
June, went away.   She is an extremely pleasant
girl, lively, cheerful, kind-hearted and good-
humoured, with a very good understanding too.

We visited Sir John Bell, shopped, and drove in
the park.

————

A splendid day, very hot.

We visited the Leopold Powyses, and afterwards
went to the garden party at Holland House, a very
pretty scene, and the fine old trees, the shady walks,
and bright verdure of the grass were very refreshing.
Those grounds are so *country-like*, it is always
difficult to believe one's-self in London, for
Kensington is really in London.   This was a
brilliant day, intensely hot.   Summer seems to have
come at last quite suddenly.

At the Holland House party we met Lord and
Lady Lilford, as well as Leopold and Mary Powys
and Minnie.

Lord Lilford is just come back from the
Mediterranean, where he has been yachting and

ornithologizing. On a small rocky island near Sardinia, he procured several specimens, both living and dead, and also eggs of a very rare species of Gull, the Larus Audouinii (if I have not mistaken the name*), which was scarcely known before.

We dined with the Aberdares:—met some remarkable men—Mr. Browning, the poet, (who is very pleasant), Mr. Goschen and Mr. Forster (ex-Ministers), Mr. George Trevelyan. I sat by Mrs. Forster, who is a daughter of Dr. Arnold, and a very clever and interesting person. Talking of Froude's book on " The English in Ireland," she said, she thought that there was some excuse for the one-sided anti-Irish violence of his work, for the Repealers and Home Rulers have held forth so noisily and so pertinaciously on the wrongs and grievances of Ireland, and on English oppression, that it was allowable to show that there was another side to the picture, and that the Irish had not been merely oppressed angels. Mrs. Forster told me that her husband and Sir Arthur Helps were the heroes of the adventure recorded by Sir Arthur in "Animals and their Masters,"† when, though very much afraid of being too late for a Council, they would not allow the cab-horse to be overdriven.

Friday, 10th.

We went out to Rose Bank and spent the evening very pleasantly at the MacMurdos. William Napier and Susie Bunbury there.

* Correct : see *Nature*. July 16th, 1874.

† Helps' " Animals and their Masters," pp. 160, 161.

Extremely fine and hot.

Fanny went to the Temple Church. I met her afterwards at the Charles Hoares. And we had luncheon with dear Kate and her husband, and spent some time very agreeably with them.

————

July 13th.

We went to the St. James's theatre. A French play by a French company (morning performance); met Kate Hoare there by appointment — saw "Andromaque."

————

July 14th.

Excessively hot.

M. and Madame Earnest de Bunsen came to luncheon.

————

July 22nd, Wednesday.

We returned yesterday to town, having been out for a week on visits; the weather extremely bright and hot all the time.

On the 14th, we went down to Lynwood near Sunningdale, the Frederick Greys'. Lynwood is a beautiful spot, and its freshness and flowers, its heath, and the smell of the fir trees, and the view over the distant country to the chalk downs, — all were delightfully refreshing in the sultry weather. We spent four days there—the 14th to the 18th, very agreeably.

The 15th and 16th, Lady Grey took us out 1874. driving in their waggonnette, through the pretty country.

On the 15th we called on Mrs. Seymour Bathurst.*

The drive on the 16th was extremely beautiful :—through a part of the woods of Windsor Park, past a number of the finest and grandest cedars I have ever seen—especially remarkable for the height and grandeur of their stems,—then winding for a long way round the shores of Virginia Water, and afterwards through a higher part of the park, to where we looked down to the Castle ;—this however was rendered rather indistinct by a hot haze. The whole drive gave us a rich enjoyment of beautiful and interesting scenery.

About Lynwood, Erica cinerea and Tetralix already in full blossom ; Calluna vulgaris not yet. Abundance of a Hieracium in flower—qu. umbellatum or boreale ? Epilobium angustifolium, growing among the heath, *apparently* wild, but I suspect originally escaped from cultivation.

The 17th, we drove (through Ascot and Bracknell) to Marchfield (Henry's house, which he bought last year), and spent the afternoon with him and Cissy and three of their children (Emily, George, and William), returning to Lynwood to dinner. Henry, I grieve to say, is again in a very distressing state of health.

While we were staying with the Greys, I renewed acquaintance with Edward Terrick Hamilton, whom

* Mother of Lord Bathurst.

1874. I had known at Cambridge, and had hardly once seen
since we were both undergraduates.   He is now the
Greys' nearest neighbour.   He was very hearty and
friendly in his manner to me.

———

July 18th.

My dear wife's 60th birthday.

We left the Greys in the afternoon, and had a
very pleasant drive to the Woking Station, about 7
or 8 miles, passing by the heathy hills where a sham
fight was performed under the eyes of the Queen,
a few days before our arrival at Lynwood.   I take
particular pleasure in that open, breezy, heath-
covered country, unencumbered by enclosures or
crops, of which there is such an extent in this part
of England.   In summer, at least, there is
something peculiarly exhilarating in it.

From Woking we went by railway (a very short
journey) to Guildford, and thence in a fly to the
Godwin Austens, at Shalford, where we remained
till the 21st.

Shalford House is between 1 and 2 miles S. of
Guildford, lying low, amidst rich moist meadows on
the banks of the Wey.   The house is of a good size,
of red brick, partly old, with some modern
additions.

Looking northward and N. E. we see the steep
face of the chalk range, the North Downs, running
eastward from Guildford, with some very large pits
exhibiting conspicuous cliffs of chalk, and nearer,
between this and the meadows, a bold hill, partly

wooded, showing at its extremity an escarpment of 1874. tawny, brown rock, belonging to the lower green sand. Austen told me that this part of the valley of the Wey, in which Shalford is situated, is a valley of denudation, the meadows along the bottom resting on the Weald clay, the overlying rocks having been scooped down to it; so that it is bounded on one side by the successive ridges of the green sands and chalk, and on the other by the lower greensand alone.

Behind the house is a pleasant, green lawn, shadowed by great old Scotch firs, on the margin of the slow, quiet, winding, little stream of the Wey, and here, after the great heats of these two days, 19th and 20th, we lounged and talked away the cooler hours of the afternoon very pleasantly.

The party in the house were Mr. and Mrs. Godwin-Austen, their eldest son Major Haversham Austen, and a younger son; a married daughter, and three unmarried daughters; the Douglas Galtons and their eldest daughter; and ourselves. Major Austen, who is well known as a geographer and a naturalist, and the explorer of some of the wildest and most difficult regions of the Himalaya, is a very interesting young man, unassuming and of great abilities and acquirements. He has a great talent for drawing, and his numerous sketches taken in various parts of India and in Burmah, are of great interest. Zoology appears to be his favourite study, and he is now engaged (as I learn) jointly with Lord Walden* in a great work on the Birds of India.

* Son of Lord Tweedale.

1874.   Mrs. Austen took us into Guildford—we spent a
long time in an old furniture shop.

Colonel* and Mrs. Lane Fox, Mr. and Mrs.
Lushington, Mr. and Mrs. Mangles, came to dinner.

———

July 21st.

We left Shalford, and had a very agreeable
journey to London, going round by Dorking and
Leatherhead, and arriving at the Victoria station a
little before 2.

From Shalford to Dorking the railway runs E.,
parallel to the chalk range and nearly at the foot of
it, through a pleasant and pretty valley. At Dorking
we changed to another line of railway, and waited
half an hour at the station, enjoying a beautiful
view of Box Hill and the remarkable gap in the
chalk range.

A beautiful and interesting scene indeed ; on the
E., the high, bold, steep front (almost escarpment)
of Box Hill, partly shaggy with wood, partly showing the bare chalk, or only sprinkled with bushes :
on the W., a beautiful hill less high and steep than
the one opposite, yet considerable, with the house
of Denbighs near the top, embosomed in wood, and
overlooking a fine, grassy slope.   A rich valley
between, and in the background to the N., Norbury
Park and House.   (This is the scene of the
imaginary " Battle of Dorking," and the ground is
admirably well described in that pamphlet by
Colonel Chesney).

* Afterwards Major Pitt Rivers.

From thence northward by Leatherhead, Epsom, 1874.
Cheam, Mitcham, Streatham, and so on over the
Battersea bridge to Victoria;—the country very
pretty, at least as far as Epsom.

Wednesday, 22nd July.

48, Eaton Place. The Charles Hoares, Lady
Bell and Edward dined with us—all very pleasant.

Thursday 23rd July.

Good news of the Kingsleys.

We went to the Zoological Gardens; saw that
very curious bird the Darter (Plotus Anhinga) from
the Amazonian region of Brazil; the first I ever saw
alive. The extraordinary length, slenderness and
flexibility of its neck, which wriggled like a snake as
the bird swam; its very small head scarcely thicker
than the beak; the swiftness with which it swam,
and turned, entirely under water, in its tank; the
glossy, rigid appearance of its plumage; the attitude
in which it sat on the rock-work after coming out of
the tank, with wings outspread, and its large tail
expanded like a fan; all very remarkable. It is
indeed a curious and interesting creature.

We dined with the Charles Hoares in Fleet
Street; met the Locke Kings, Edward, Mr. and
Mrs. Paget. Kate Hoare charming.

The Darter is well described (under the article
Pelecanidæ) in the *Penny Cyclopædia*. It seems
to be new to the Zoological Gardens, for it is not
noticed in the catalogue for 1872.

1874.   Friday, 24th July.

Mr. Dyer came to luncheon.   We drove a long way to the N.W., quite out of town, to call on Sir Charles Murray, who has bought and furnished a large house there.   It is quite a sight to see, for the wonderful wealth of decorative art, with which Sir Charles and Lady Murray have enriched it.   It is a perfect museum of splendid ornamental furniture, hangings, china, &c.—but not at all what I can attempt to describe.

In our way back from Sir Charles Murray's, we called on Miss Gordon, the daughter of Sir Willoughby Gordon.   She showed us a very lovely portrait, said to be of the famous Lady Coventry by *Henry Morland*, and very like a picture which I noticed in the portrait exhibition of 1867, but I think Miss Gordon's picture is even more beautiful than the one in the exhibition.

Saturday 25th July.

Down to Barton.

11.30. train from St. Pancras, arrived safe and well—thank God.   We brought Louisa MacMurdo down with us—great crowd at Cambridge ; we had the company of Mrs and Miss Horton in the carriage from Cambridge.   All well at home—thank God.

Sunday, 26th July.

I will mention some other pictures in the Royal Academy, which, besides those set down under June 4, I noted with approbation in subsequent visits :—

"Little Foxes," by *S. Carter*, a charming little 1874.
picture of a brood of young foxes looking out of a
hole in a bank, most bright and life-like. "The
Picture Gallery," by *Alma Tadema*—ancient Roman
connoisseurs scrutinizing a picture, very clever.
"Crossing the Heath," by *Eddis*, extremely pretty.
"A Sunny Summer Evening in the Meadows," and
"Fording a Brook in the Marshes," both by *Cooper*,
and both delightful. I should still rank *Millais's*
"North-West Passage" above all other pictures in
this Exhibition.

Wednesday, 29th July.

We drove out with Louisa MacMurdo to the
tumuli on the Rougham Road.

Thursday, 30th July.

The Assizes at Bury: I was foreman of the
Grand Jury; dined with the Judges, Chief Baron
Kelly* and Justice Keating: business of the Assizes
finished in the one day.

Friday, 31st July.

Mr. and Mrs. Abraham and their daughter
Louisa came to luncheon :—afterwards we went
with them to the Wilsons at Stowlangtoft, where
was a Garden Cottage Show and a great many
people.

---

* Sir Fitzroy Kelly.

1874.   **August 1st.**

Sir Charles' Journal this month begins with the following quotation :—

" I will chide no breather in the world, but myself, against whom I know most faults."

LETTER.

Barton,
August 1st, 1874.

My Dearest Cissy,

It makes me very unhappy, more so than I can express, to hear of Henry suffering so terribly, and I cannot help feeling very anxious as you may well suppose. No doubt all is done for him that human skill can do, and one can only hope and trust in God and pray that he may bless the means that are used. I cannot express how much I admire Henry's fortitude and patience, or how much I feel for you both. May God bless and comfort you.

It is a great comfort that Sarah Craig is with you.

Ever your most affectionate brother,

CHARLES J. F. BUNBURY.

JOURNAL.

Sunday, August 2nd.

At home.

We went to morning Church and received the Communion.

Mr. and Mrs. Percy Smith came to tea.

Monday, August 3rd.

Mrs. Wallis died this morning.*

The Temperance Society of Bury held their annual festival in our grounds—a great assemblage and much merry-making, with perfectly good behaviour.

Tuesday, 4th.

Fanny with the two MacMurdo children (Louisa and Arthur), went to Mildenhall and returned to dinner.

Thursday, August 6th.

We went to Hardwick in the afternoon, saw Lady Cullum, and stayed some time with her : met Mrs. William Blake with her three girls, who are very pretty, also the three youngest Miss Floods.

Friday, 7th.

Mrs. Byrne (Fanny's Aunt), arrived.

Monday, 10th.

Talk with Scott about Mildenhall Church matters.

Tuesday, 11th.

Leonora and her husband arrived.

Wednesday, 12th..

Took a walk with Leonora.

* She was the widow of our coachman, who had died in 1866, and who had lived with Sir Henry and Sir Charles Bunbury since 1824.—(F. J. B.)

We have been quite quiet and happily busy since our return from London. Fanny much engaged with Arthur and Louisa MacMurdo.

On the 6th I was delighted by seeing in the newspapers an announcement of Kingsley's safe arrival at Liverpool: I wrote immediately to Rose, to welcome them home, and on the 11th I received from her a charming letter, from Eversley, whither they had returned safe and well. She says that her father had had a serious illness—an attack of pleurisy, on the way back from California, but that he is now well, though still weak.

_____

We had a great alarm about Henry. Late in the afternoon of Friday the 14th, we received an alarming telegram from Cecilia, saying that he was much worse. We immediately made our arrangements for going up to London on the 15th; but in the forenoon of that day, a little before the time when we were to have set out, came a telegram with a much more comforting account; and another in the afternoon was still more satisfactory, saying that he was "out of all present danger." We have therefore remained at home; and the subsequent accounts have all been favourable.

Susan Horner arrived to-day.

_____

Mr. Buttery arrived to examine our damaged

pictures with a view to repair, and repaired one of them very well.

Songs and merriment in the evening

———

Began to revise my collection of fossil plants, comparing them with Schimper's Paléontologie Végétale.

Read over Charles Lyell's old paper on Geology of Forfarshire.

———

A brilliant day. Fanny and Katharine, with all the young people, went to Ely and spent the day there, returning late.

Arranged some minerals and went on with revision of fossil plants.

———

Went to morning Church. An excellent sermon from Mr. Percy Smith.

Read Kingsley's fine sermon on God and Mammon.

———

Brilliant and very dry weather, with cold east wind. Long talk with Scott on the scarcity of water.

———

Dear Arthur went back to school.

1874.   Monday, 31st August.

Leonora and her husband and children went
away.

———————

Saturday, 5th September.

Leonard Lyell and his wife arrived.

Lady Cullum, the Wilsons (4), Mr. and Mrs.
Bland, Mr. and Miss Bevan, Mr. Hall, and young
Flood, dined with us.

———————

Tuesday, 8th September.

Since the 7th of last month, we have had the
house filled with a family party. Katharine and her
husband and daughter, with Leonora and her two
daughters, Annie and Dora, came on the 15th
of August; Susan (Horner) on the 17th; Leonard
Lyell and his bride on the 5th of this month. The
Pertzes went away on the 31st of August Katharine,
Harry and Rosamond yesterday.

Mrs. Byrne and Susan and the Leonard Lyells
remain.

I have had much enjoyment in botanical conver-
srtion with Katharine, and in looking over dried
plants with her.

———————

Saturday, 12th September.

The Leonard Lyells went away on the 10th.
Mrs. Byrne on the 11th. Susan Horner alone
remaining with us. Leonard's wife is a very intelligent, well-educated, pleasant young woman. I

think his marriage will do him good in various ways. 1874.
He has considerable abilities, and is very estimable.

_____

Mrs. and Miss Anderson came to luncheon, and
walked with us afterwards.

_____

A beautiful day. My Barton rent audit, and
luncheon afterwards with my tenants—very satisfactory and comfortable.

We walked about the grounds and put labels
to several of the trees.

_____

Up to London with Susan Horner and Louisa
MacMurdo. Susan went to stay with Charles Lyell
and his sister.

_____

My nephew Cecil, with his extremely pretty and
interesting young wife (Susan Napier) and Charles
Napier, came to us on the 16th. Kate and Charles
Hoare and his brother William, on the 17th, Minnie
Boileau on the 19th. These with Susan Horner
made an exceedingly pleasant party.

Susie Bunbury is looking well in both senses ;
for without losing any of her delicate beauty, she
has a less fragile, shadowy appearance than she
used to have.

Wednesday, 23rd September.

Called on Lady Mary Egerton.   A visit from
Susan MacMurdo.   Then we went to the Cloisters
at Westminster to see the dear Kingsleys—saw Mrs.
Kingsley, Rose and Mary—delightful.

Thursday 24th September.

Received at last a letter from Edward from
Stralsund—sent it to Henry.

Susan Horner came to luncheon, and Fanny
went with her to the British Museum.

We dined with the Charles Hoares, and met the
Bishop of Bath and Wells.

Yesterday we had the great pleasure of seeing the
Kingsleys — Mrs. Kingsley, Rose and Mary. —
Charles Kingsley was resting.   I grieve to hear
that he is in weak health, having had a severe
attack of illness (congestion of the liver) since they
came to Westminster, though he has recovered
perfectly (they say), from the dangerous attack of
pleurisy with which he was seized in California.
That illness must have been a terrible trial to poor
Rose, and her face shews the effects of harassing
anxiety: she looks much worn.   Dear Mrs.
Kingsley, too, is looking ill, — Mary blooming.
They were as cordial and charming as possible.

-- ----

Friday, 25th September.

Rose Kingsley came to us about noon, and

brought a part of her collection of dried plants from North America. She has most kindly and generously given me her whole collection made in this last journey, principally in Colorado and California.

I looked through one large package with her, and found them extremely interesting: — splendid Astragali, Pentstemons, Gilias, Mentzelias, some exquisite Ferns and Grasses, and a great many which I did not know. It was very pleasant to look over them with her, and to hear her animated descriptions of the localities in which she had collected them. Her enthusiastic nature and great powers of observation and expression (like her father's), make her conversation very interesting.

Charles Lyell and his sister, Kate Hoare and Susie Bunbury came to luncheon, and we had a very pleasant little party and very lively talk.

Afterwards the Lyells and Kate went away, and about 4 p.m., we with Rose, Susie and Louisa Mac Murdo proceeded to Westminster Abbey, entering it from the Cloisters. We were joined by Susan MacMurdo, and for a little while by Charles Kingsley, whom I was delighted to see, now for the first time since his return to England. He reports himself well, but I am afraid is far from strong.

We remained nearly two hours in the Abbey, and saw all that was most interesting in it, under the guidance of Rose, and an admirable guide she was.

The Abbey is indescribably interesting: the resting-place of so many famous and extraordinary men and women, and containing the memorials of

1874. so many great events. The magnificent shrine of
Edward the Confessor, erected by Henry III.; the
huge, grim, massy, unadorned monument of Edward
the First; the gorgeous tomb of Henry VII., with
its very beautiful sculpture, the work of the Italian,
Torrigiano; the tomb of Henry's mother, Margaret
of Richmond, also by Torrigiano (the reclining
figure remarkably beautiful); that of Sir Francis
Vere, with the four kneeling knights supporting over
him the canopy on which his arms are laid (this is
a noble monument of a very brilliant chivalrous
character); the monument (perhaps even finer than
Vere's), of Sir — Norreys? supported by his six
sons, of whom all but one died in battle; that of
Sir Robert Walpole's first wife, a very beautiful,
upright, single figure, closely imitated from the
famous antique Pudicitia in the Vatican; the
monumental statue of James Watt, most noble,
almost sublime in its severe simplicity; and that of
Francis Horner, also very noble; these were the
particular objects which made most impression on
me.   But, in truth, I saw in this visit more interesting objects than I can well remember, and received
more ideas than I could well arrange or retain in my
head.   More recent memorials in the Abbey :—bust
of Mr. Maurice, like, but too stern, deficient in his
peculiar sweetness; bust of Thackeray; of Dickens
there is nothing but his name, on a slab in the
pavement (this by his own desire).

Add :— the two Coronation Chairs, which reminded me of Sir Roger de Coverley's visit to the
Abbey

Soon after our return, I read over those two 1874.
capital papers of Addison in the *Spectator*.

September 26th.

A delightful conversation with Charles Kingsley,
who came to see us after the Abbey service, sat
some time with us, and then I walked back with
him, through St. James's Park, nearly to Westminster. He talked admirably.

Fanny having given the conversation a serious
turn, Kingsley spoke beautifully on the recognition
of friends by one another in the future life, and on
different grades of elevation in that state—warmly
supporting the affirmative as to both. One expression he used I particularly remember : " A man who
has found a woman his guardian angel in this life,
may be allowed to hope that she will be so likewise
in the life to come."

He said also, that he did not believe that God
would have given us the deep, instinctive longing
for re-union with our friends in a future state any
more than the instinctive longing for immortality,
if He had not intended that those aspirations should
be satisfied.

Kingsley's illness, while in America, was brought
on by the excessively rapid and violent changes of
temperature in passing from the hot plains so
rapidly as is now done, to the icy high levels. For
instance leaving St. Louis in absolutely tropical
heat, he and Rose, in two day's time, found themselves passing through ten feet of snow on the
Rocky Mountains.

1874.   He describes the Central parts of North America
from the lower Missouri, westward almost to the
mountains of California, as generally a dreary, bare,
desolate wilderness, almost treeless, with a climate
which he thought detestable, dry to excess, with the
fiercest extremes of heat and cold ; scarcely ever
any rain. Vast, monotonous plains, almost destitute
of vegetation, or covered with a low, monotonous,
grey under-shrub, the Artemisia tridentata *(sage-brush)*.

The soil however, in general not barren in
itself, where it can be watered ; hence the fertility of
the tract about Salt Lake City, to which extreme
industry has brought down abundance of water from
the mountains. There are some wide tracts of
absolute sandy desert.

Pike's Peak and several other peaks in the Rocky
Mountains, in Colorado, are (Kingsley says)
between 14,000 and 15,000 feet high ; yet no continuous perpetual snow, only patches and streaks.

One mountain called the Holy Cross Mountain,
has a singular arrangement of snow in gullies on its
face, forming the perfect figure of a cross.

The flowers of the Rocky Mountains in spring and
early summer, certainly beautiful, yet he thought
them not equal in their general effect to those of the
Alps and Pyrenees.

The Yosemite Valley, Kingsley says, did not dis-
appoint him at all; it is quite as wonderful and
as beautiful as it has been described. Yet it is
difficult to take into the mind a true measure of the
enormous size of the different objects, because *all*

are so symetrically vast, and also because of the 1874.
uncommon clearness of the air.—There are not only
the famous big trees, but a general prevalence of
truly gigantic coniferous trees : — Pinus ponderosa,
nobilis, and (grandest of all next to the Welling-
tonias) the P. Lambertiana, called in that country
the Sugar Pine, because a truly saccharine juice
is secreted beneath its bark.

The famous cañons (pronounced *canyons*) of those
Western regions, Kingsley does not doubt, have
been formed entirely by means of erosion by water
—"sawn out," as he expressed it.

_____

September 27th.

We went to the Kingsleys, and with them to
afternoon service in the Abbey, where Kingsley
preached a beautiful, noble sermon, on a fine text of
the 3rd chapter of St. Peter, verse 8-12.— "*Finally,
be ye all of one mind, having compassion one of another.*"
—and so on. Being placed however at some distance, I had some difficulty in hearing him, and
lost some portions of his discourse.

_____

Monday, 28th.

Walked out with Fanny.

We called on the Benthams, and saw them.

Minnie Boileau came to stay with us for one
night.

_____

P

Mr. Bentham breakfasted with us, and was very
agreeable; told us much of John Stuart Mill, whom
he knew intimately in his youth, though in their
latter years, owing to difference of habits and pursuits, and opposition of opinions, they saw little
of each other. Old Mill (James) appears to have
been a still more odious man than I had supposed.
His wife was a weak, insignificant, poor creature,
and Mr. Bentham says he treated her brutally, and
not only ill-treated her himself, but actually en-
couraged her own children to ill-use her. While he
crammed his eldest son with knowledge, in the
extraordinary way we have read in the autobiography, he left his other children in total neglect;
if John Mill had not himself taught his brothers
and sisters, they would have been utterly uneduca-
ted. Bentham said he remembered J. S. Mill when
a child of seven or eight years old, engaging in
an animated argument with Lady Spencer on the
comparative merits of Marlborough and Wellington.

Mr. Bentham said, that in John Mill, with all his
errors and eccentricities and offensive notions, there
was much that was good, and even fine; this I had
already concluded from his autobiography. He
suffered from want of sufficient intercourse with
others. In youth his manners and behaviour were
indescribably awkward and uncouth.

James Mill was ungrateful to Jeremy Bentham, in
whose house he had lived for years. Anecdote
of his extraordinary wholesale borrowing of books
from Jeremy Bentham, to the amount at last of

a thousand volumes. George Bentham also talked 1874.
well on the light literature of France and Germany.

Bentham told me he was now working at the
Asclepiadeæ, for the Genera Plantarum — a peculiarly difficult family of plants. The Compositæ he
said, are difficult only on account of their prodigious
number, and the unskilful way in which they have
been treated by many authors. Their flowers are
easily dissected and examined, and the important
characters are well preserved in the dried state.

We had a very pleasant little dinner-party :—the
Bishop of Bath and Wells, Kate Hoare (her
husband was kept away by business), Rose Kingsley,
the MacMurdos and their daughter Mimi, and Mr.
Bentham ; all very agreeable. Lady Head and her
daughter in the evening.

October 1st.

Susan Horner set out very early on her long
return journey to Berlin and Florence. She has
been a most pleasant member of our society ever
since the 17th of August. She is looking remarkably well, seems to be in excellent spirits, and is
extremely agreeable.

Mr. and Mrs. Kingsley and Kate Hoare came to
luncheon : Rose and Mary (by a mistake), not till
afterwards. All were very agreeable.

Kingsley said he thought it would be absolutely
necessary for the United States Government, within
a few years, to take possession of the Sandwich
Islands, and for our Government to occupy the

1874. Feejee Islands and perhaps some others, in order to prevent the Pacific from becoming the prey of pirate communities.

The South Sea islands led to the subject of cannibalism. Kingsley said that his brother, when with Lord Pembroke, saw abundant evidence of its prevalence in some of those islands. He thought that the inclination to it might sometimes arise, not from absolute want of any food, but from the desire for animal food, which might be scarce in some islands.

Kingsley's amusing accounts of the coolness and disagreeable precocity of children in America.

Late in the afternoon we visited Charles Lyell, who seems well and cheerful.

October 2nd, Friday.

Mary (Leonard) Lyell came to luncheon with us, and brought us the first news of the tremendous explosion of gunpowder on board a barge on the Regents Canal, by which all that neighbourhood was terrified, and immense mischief done :—many houses ruined, and a vast number damaged. Happily, Katharine has not suffered, beyond the breaking of a few windows. Here we slept through it all, and heard nothing. The house of M. Alma Tadema, the artist, is said to have been almost completely ruined.

Mr. Walrond dined with us.

Saturday, 3rd October.

Fanny went with Susan MacMurdo to Hertford,

to see Arthur at his school: they returned about six.

I went to visit Lady Bell, as from the situation of her house in Albany Street, and her great age, we feared that she might have suffered some hurt from the explosion. But I found her very well and in very good spirits. She said, however, that the noise and shock were tremendous, and like, what she imagined, the sensation might have been if heavy canon had been discharged against the house. Yet, strange to say, not even the windows were broken. Katharine came in while I was there, and confirmed what I had heard before as to the immunity of *her* house from serious damage, while in the house next door plate glass windows were shattered.

After my visit to Lady Bell, I drove along the road skirting the canal, round the north end of the Park, and saw long ranges of houses with their windows totally destroyed: in many cases the window-frames and shutters demolished as well as the glass.

I met at Lady Bell's Mrs. Grey and Miss Sherrif.

———

Monday, October 5th.

We went down to Marchfield, to see Henry and Cissy, and spent the night there.

I cannot dwell on the particulars of poor Henry's sad and suffering state. The only bright side of this melancholy state, is, his amazing courage and patience and unselfishness, always anxious that his sufferings should not destroy the cheerfulness of

1874. those about him. He has all the comfort that a
devoted wife and a bright, cheerful, loving daughter
can give him.

On the 6th we drove in a hack carriage from
Marchfield to Windsor, passing through a beautiful
country, but the day was so bad that we could not
enjoy it much.

At Windsor we spent a few hours very pleasantly
with dear Minnie and Sarah and Albert Seymour:
they are living there for the present, as Albert is on
duty. He went out walking with me in the interval
between two storms of rain, and showed me the
famous Terrace of the Castle, which I had never
seen before, though I had so often heard of it. The
dark and foggy weather prevented us from seeing
far, but I could form some idea of the magnificence
of the view under favourable circumstances, and it
was interesting to have even a birds'-eye view of
Eton, Datchet, Mead, etc.

We returned to Town by the Great Western line,
to Paddington. That same evening (the 6th)
Edward came in, just returned to England, very
much fatigued, having had a very bad passage from
Ostend, and having been without food for 12 hours.

He has been making a tour in Sweden, and a
letter which he wrote to me from Stockholm not
having reached me, we were a good while without
news of him.

––––––

Wednesday, October 7th.

Susan and Mimi MacMurdo, William Napier,

Edward and Clement came to luncheon, and the 1874. three latter dined with us.

We visited Sir W. Boxall, went also to Week's to enquire about glass-houses.

Edward tells me, that he found the Swedes the cleanest and the most honest people he ever was among. He enjoyed his tour much.

The country is not grand, but much of it very pretty and picturesque :—innumerable lakes and rivers, much wood, moderate hills, with picturesque promontories and tongues of land running into the lakes. About Upsal, however, the county is decidedly ugly, tame, flat, dreary moorlands. The famous iron mines of Dannemora, also, are situated in a dead flat.

October 8th, Thursday.

We saw Gustave Doré's picture, "The Dream of Pilate's Wife," very striking, impressive and powerful.

Friday, 9th October.

Much shocked by hearing (yesterday evening) of the death of Mr. Twistleton. We have heard no particulars yet, but it would seem that it must have been rather sudden. He was a man to be much missed and regretted.

Susan MacMurdo, Kate Hoare, Katharine and Rosamond, W. Napier, Cecil and Clement came to luncheon : C. Lyell and his sister and Edward afterwards.

October 10th.

Down to Barton by the 11.45 train from Bishops-
gate: Emily MacMurdo and Clement with us. We
arrived safe and well, and found all well at home—
thank God.

———

Monday, 12th October.

Wrote to Lady Grey.

Edward's missing letter, from Stockholm, written
August 27th, has at last arrived, it had been
insufficiently directed. In it he says:—

"Stockholm is a very picturesque and peculiar
"place, and would be still more striking if the
"buildings were finer or more picturesque; but in
"this respect it is very inferior to the southern cities
"—and the few buildings there are especially the
"Royal Palace which is twice as big as the Pitti, but
"has no other merit, only serve to crush and dwarf
"the rest of the town. Still, the combination of rock,
"wood and water, with a large town and quantities of
"shipping, is very picturesque, and there is great
"variety, from the number of branches and arms of
"the lake by which it is intersected in different
"directions. The communication is kept up in a
"great degree by a number of very tiny steamers,
"which are continually going from one quarter of
"the town to another."

Of Upsal (where he saw the tomb of Linnæus
and that of Gustavus Vasa), he says:—

"There are several things worth seeing at
"Upsala, but the country around is very un-

" interesting. It is, indeed, the only part of 1874.
" Sweden that I should call ugly : all the rest is very
" peculiar, and without ever rising to anything
" like fine scenery, has a charm of its own, from the
" perpetual combinations of rock, wood and water,
" producing endless variety in detail, though con-
" siderable monotony in the *tout ensemble.*"

Wednesday, 14th October.

Began to read my father's Manuscript of
Castruccio.

October 15th, Thursday.

Rose Kingsley's collection of dried plants from
N. W. America (see September 25th), having
arrived, I began to examine and arrange it, and
wrote to her on the subject. It is a most friendly
and generous gift to me.

Saturday, 17th October.

We spent the afternoon (which was wet) in
arranging and cataloguing our new books.
Went on with the arrangement of Rose Kingsley's
dried plants.

Monday, 19th October.

Took a walk with Minnie and Edward.
The price of wheat in Bury market (Scott tells
me), is at present not more than 21s. per *comb* (42s.

1874. per quarter). That of barley from 18s. to 25s. per
comb (36s. to 50s. per quarter). The ordinary
wages of farm labourers in this parish, which had
been raised in the summer to 14s. per week, is now
reduced to 13s.

_____

Tuesday, 20th October.

Mr. Bevan and two of his daughters dined
with us.

_____

Wednesday, 21st October.

I walked about the grounds with Susan Mac-
Murdo, Lady Head and Lady Rayleigh.

Miss Bucke and her niece, Mrs. Davie, came to
luncheon.

_____

Thursday, October 22nd.

A visit from Patrick Blake.

_____

Tuesday, 27th October.

Most part of the day at the Quarter Sessions at
Bury.

_____

Thursday, 29th October.

Had a very pleasant walk with Minnie Powys and
Mrs. Montgomerie. Most of the party went to the
Bury Ball; Minnie, the MacMurdos, Edward and I
stayed at home.

_____

Friday, 30th October. 1874.

Our usual dance very successful.

———

Saturday, 31st.

A succession of gay and lively company in the house for the last two weeks. The MacMurdos all the time, from the 20th October to this morning: Lady Head and her daughter Amabel and Lady Rayleigh from the 20th to the 24th: Edward Campbell and his daughter Annie and his son Guy for only two days.

Our party of last week was expressly for the Bury Ball and for Fanny's dance on the 30th; it consisted (besides ourselves, the MacMurdos and Edward), of the Barnardistons, Mr. and Mrs. Montgomerie, two Miss Newtons (aunt and niece), Rosamond Lyell, my niece Emily, Sir John Shelley, John Hervey, Mr. William Hoare and Clement.

Minnie Powys would not go to the Bury Ball, but was at our dance, and is still here: she is charming as ever. But we missed Octavia and Wilhelmina Legge, who have for several years been constant and agreeable members of our October society; and I have never grown reconciled to the absence of Sarah and Kate Hervey. The MacMurdos, as usual, very agreeable, Montague especially. The Barnardistons and Montgomeries very pleasant: Mrs. Montgomerie remarkably pretty. Sir John Shelley and W. Hoare very pleasant young men.

———

Tuesday, 3rd November.

Sir John Shelley went away, also Edward in the afternoon.

————

Thursday, November 5th.

Delightful—most delightful news, yesterday, of dear Sarah Hervey's engagement to be married. She announced it to us in a letter which was quite delightful from its overflowing happiness, as well as from its warm expressions of feeling and affection towards us. Her betrothed is Mr. Sanford, of Nynehead Court, near Wellington in Somersetshire. I have not for a long time heard anything which has made me so happy. I have again and again recorded in this journal my very high opinion of Sarah, my exceeding admiration and affection for her, and as I have no doubt that a well-assorted marriage gives the greatest amount of happiness which this life admits of, I have most earnestly wished to hear of her meeting with a worthy husband. Many are the dreams in which I have indulged, on this subject, and I began almost to fear that no man could be found, in these days, qualified to make her happy. I am sure that, to be worthy of *her*, a man must possess extraordinary merit. And, if we may judge by what she writes of him, I should hope that Mr. Sanford may really be worthy of such a blessing. Whatever his merits may be, I am sure that he is a most uncommonly fortunate man. God grant that the marriage may turn out as happy as it promises, and that they may long be preserved to each other. The only draw-

back in the case, as far as we yet know, is, that Mr. 1874.
Sanford is much older than Sarah—more than 20
years older : this is of no consequence now, but 20
years hence he may perhaps be infirm while she is
still in middle life.

Dear Minnie Powys went away.

Two Miss Newtons, aunt and niece, arrived, and
Emily went with them to the Thetford Ball.

Saturday, 7th.

Received a charming letter from Sarah Hervey.
Took a walk with Fanny and her dogs. Emily
went away.

Tuesday, 10th November.

Received a very pleasant, warm and cordial
letter from the Bishop of Bath and Wells, to whom
I had written to congratulate him : he is delighted
with the marriage, and says that " we have every
ground of confidence that, by God's blessing, Sarah
will have a very happy future before her." He
feels, of course, that to him and her mother, the
loss of her society will be " incalculable," but it is a
comfort that her home will still be in the same
county with them.

Barton, Bury St. Edmund's,
11th November, '74.

*Extract of a letter from Sir Charles Bunbury to
Mrs. Katharine Lyell.*

" I am reading Sir Erskine May's 'Constitutional

1874. " History of England,' which is very good as far as
"I have gone; he takes up the history where
" Hallam leaves off,—at the accession of George the
" Third.—

   " Reading also Mr. Charles Greville's Journal,
" just published, there is a great deal that is curious
" and interesting, but I think it is odd that they
" should have been published so soon, for they
" contain the most unceremonious remarks, if not
" on people now living, yet on their fathers. They
" give me a still worse opinion of George the Fourth
" than I ever had before.

   " I am busy arranging Rose Kingsley's collection
" of dried plants from Colorado, which is a rich and
" interesting collection. I think I told you that she
" had most generously given me the whole of it.
" There are a great many fine things : splendid
" Pentstemons, Pedicularises, Polemoniaceæ, Lupins,
" Astragale, Loaseæ, and many others ; few Ferns,
" but some exquisite little Notholœnas, which I
" daresay you have had from Mr. Redfield.

   " I am also going on with the revision of my fossil
" plants, noting them according to Schimper."

———

Thursday, 12th November.

There is another piece of good news which has
made us happy : we had a telegram yesterday,
telling us that dear Sarah Seymour became a
mother that morning, and that all was well so far.
To-day Fanny has had a letter from dear Minnie ;
mother and child doing well, and every reason

to hope for a continued good report. The child is a 1874.
boy.

_____

We have been much shocked by the news of the sudden death of Mrs. Hooker. It is a terrible blow, and a most grievous loss to her poor husband, who was absent in London at the time, having left her that morning apparently in good health; and returned to find her dead.*

_____

Meeting of magistrates at Bury, to elect governor of gaol; nine candidates put to the vote—Captain Twyford elected.

Mr. and Mrs. Goodlake, Susan Hervey and her brother Dudley arrived.

_____

Mr. and Mrs. Robeson came to luncheon.

_____

Fanny went with several of the party to see Hardwick. I walked with Lord and Lady Rayleigh.

Lady Cullum and the Victor Paleys dined with us.

_____

* The cause of Mrs. Hooker's death, we have since heard, was the rupture of a blood vessel near the heart.

1874.   Saturday, 21st.

Received a very pleasant letter from Albert
Seymour, giving an excellent account of Sarah
and the baby, and in the kindest manner asking
me to be one of the godfathers, which, of course,
I am very happy to be.

———

Sunday, 22nd.

A remarkably agreeable company in our house
these last ten days:—Mrs. Ellice and Helen from
the 11th to the 14th: Harry Bruce (Aberdare's
eldest son), from the 9th to the 20th: Susan
Hervey (Lord Charles' eldest daughter), and her
brother Dudley from the 16th to the 21st: Mr.
and Mrs. Clements Markham and Lord and Lady
Rayleigh from the 17th to the 20th: Mr. and
Mrs. Goodlake (*she* was Cissy Ellice), came on the
16th and are here still.

Mr. Clements Markham is a remarkably clever
and agreeable man, has travelled a great deal, is
generally well cultivated, and has a great store of
knowledge on all matters connected with geography,
—communicates his knowledge readily and agreeably. His travels and adventures in the Andes,
his successful exertions in procuring the valuable
Chinchona plants of that country, and introducing
them into India, are celebrated.

Mrs. Markham is very clever, lively and pleasant.
Though a little, slight, delicate-looking woman, she
accompanied her husband in his adventurous and
dangerous journeys in the Andes. She admits that

this was hard work,—that though she is glad to 1874. have seen that country, she would not wish to repeat the enterprise; but their travels in India were really enjoyable.

The two young Herveys, Susan and Dudley, both very interesting, both have the refinement characteristic of the Herveys. Susan has not the animation of her charming cousins, but has a very well cultivated mind, is thoughtful, earnest and very well read; her conversation very good. Her brother is a remarkably pleasing and interesting young man; he has been in some office at Singapore, from which he returned to England on account of his health early in this year; his manners are peculiarly pleasing, and he is extremely intelligent, earnest, thoughtful, observant, (indeed with considerable power of observation, I think), and very desirous of knowledge.

————

November 24th.

Harry Bruce also is a very pleasing, *likeable* and interesting young man, very well informed for his age, observant and fond of knowledge. Having delicate health, and being the eldest son, he has no profession.

Our party was a remarkably *travelled* one :—

Mr. Markham is a celebrated traveller, and Mrs. Markham was his companion in some of his adventurous journeys; Mr. Goodlake has been almost everywhere; Harry Bruce and Lord Rayleigh up the Nile; and Dudley Hervey in the Malay Islands.

1874.   Mr. and Mrs. Goodlake went away in the afternoon ; and we were at last alone.

Took a walk with Fanny and our little dog.

My leisure time has been much occupied in the
arrangement of the very interesting collection of
dried plants from Colorado, given me by Rose
Kingsley.

Weather since the 20th, frosty and very cold.

Thursday, November 26th.

Not being very well Dr. Macnab would not allow
me to go out, so Fanny went alone to dine at
Stowlangtoft.

Saturday, 28th.

In the evening resumed the reading of Dorothy
Wordsworth's Journal.

Monday, 30th.

Important business with Scott — signed agreement (drawn up by Mr. Nicholl) relating to dilapidation money for Mildenhall.

December 1st, Tuesday.

We spent an hour in re-arranging books in
porch room.

Wednesday, 2nd.

Out with Fanny and Scott inspecting and noting
trees.

Our dear, good Mrs. Stocks, formerly our house-keeper, died yesterday evening. Her departure had been expected for some time, indeed she had been in very bad health for years past, and for a good while it had been known that her malady was consumption, and that recovery was hopeless. She was an excellent woman, of a very good under-standing and judgment, and most thoroughly trustworthy; far above the common run both in good sense and moral qualities.

Received charming letters from *both* the Sarahs, and from both in acknowledgment of presents sent to them: from Sarah Hervey for a wedding present (a gold necklace and ear-rings)— and from Sarah Seymour for a christening-cup for her baby.

Saturday, 5th December.

Sarah Hervey's wedding day.

Wrote to Mr. Nicholl on business concerning the Mildenhall dilapidation money.

Mrs. Rickards sent me her collection of Sea-weeds.

December 7th.

Dear Sarah Hervey was married to Mr. Sanford on Saturday, the 5th; and this day we received a Wells paper, sent by Lady Arthur, with a very full and animated description of the wedding, which seems to have been everything that could be desired.

1874.   I can imagine how interesting and lovely she must have looked.

Lord John's speech in proposing the healths of the bride and bridegroom, was excellent.   So dear Sarah is now Mrs. Sanford, most heartily do I wish them both all possible prosperity and happiness.

But, this same day we received a very sad, distressing and alarming piece of news : that dear Mrs. Kingsley is dangerously ill.   We have been expecting a visit from Rose and Mary, and I was looking forward with eagerness to seeing them on the 8th, when a note came from Rose from Eversley, telling us that they had been re-called from Cambridge (where they had been staying) by the news of their mother's dangerous illness.   Spasms of the heart is said to be the malady.

December 11th.

Most sad news of dear Mrs. Kingsley.   Their friend and neighbour Mrs. Martineau, writes by Rose's request to tell us that the doctor gives them *no hope*—that there may be a partial and temporary rally, but there can be no recovery.*   Happily, however she is now free from pain, and is perfectly calm and composed, aware of her danger, and able to talk with them.

December 15th.

We have also been in a good deal of anxiety

* She did not die till February, 1892.

about Charles Lyell, who has had a bad fall down stairs, and cut his head very severely; a serious thing at 77; but he is doing well.

December 16th.

I am happy to say that there are now better accounts of Mrs. Kingsley; Fanny has had a postal card from Mrs. Martineau, every day since the 11th, and we have also heard through Mr. Percy Smith from the doctor at Eversley, a rather less despairing account; and now Mrs. Martineau writes that there does seem to be a real improvement, and that the doctors hope she may live some months, though her state must always be very precarious. God grant that the terrible blow of her loss may be averted from her poor husband and children.

December 21st.

I find I have quite omitted to mention our last batch of company, though it was very pleasant. The members were — Leopold and Lady Mary Powys, Sir Edward Greathead, the Louis Mallets, Lord John Hervey, and Charles Strutt.

We were disappointed of several guests whom we had expected—and above all, the absence of the poor Kingsley girls was a grievous loss. But those who did come were all pleasant. Lepold Powys is a man after my own heart, and Sir Edward Greathead is uncommonly agreeable. Louis Mallet was not at his best, being oppressed by a severe cold; but he is a man of so much knowledge,

1874. thought, and earnestness of character, that I feel it is useful to be in his company.   His wife is very agreeable ; Mary Powys most warm hearted, friendly, cheerful and lively.

LETTER.

Barton Hall, Bury St. Edmunds.
December 21st., '74.

(Extract from letter of Sir Charles Bunbury to Mrs. Katharine Lyell.)

I rejoice with you in the success of the Bishop of Natal in pleading the cause of the oppressed Zulus, and in obtaining (as I hope it will prove) from the Colonial office a redress of their wrongs.   I only hope it will not turn out (as so often in the Spanish Government of America) that local interests and passions can neutralize the efforts of the home Government in the course of justice.

The correspondence between the Bishop of Natal and the Dean of Westminster (which Fanny read to me this morning from the *Daily News*) is very honourable to both.   Fanny and I were both delighted with it ; she begs me to say she has forwarded it to-day, and that she means to get another copy of the newspaper.

Mrs. Rickards has given me all her sea weeds, including four volumes of Mrs. Wyatt's famous collection of Devonshire Algæ, all named, by Mrs. Griffiths.   We may look them over together the next time you are here.

JOURNAL.

We heard of the death of Charles Austen, who indeed, had for a good while been in feeble health. I had not seen much of him since 1868.   He was a man of great ability, and very extensive information.

———

December 26th.

Katharine sends an excellent account of Charles Lyell, who has so much recovered as to be able to go down (helped and supported but not carried) to the dining room.

Our last account of Mrs. Kingsley is, that the doctors still do not expect her to recover, but that she is free from pain and almost from uneasiness, except a fit of breathlessness for a certain time each day; that her mind seems as clear and strong as ever, even brilliantly so.

———

December 28th.

The year which is now drawing near to its close, has been one of a very " mingled web " of a mixture of joys and sorrows; or rather indeed of actual joys and threatened sorrows.   I have great cause to be deeply thankful to God, that the greatest and most precious of all my blessing is not threatened; that my beloved wife is in reasonably good health,

1874. and that we are as firmly knit in bonds of love and harmony as ever.

My dear brother Henry is indeed still alive, but lingering in the misery of an incurable and most painful disease, without hope of recovery or of even temporary amendment. How long this may last God only knows, but we could not be surprised any day if we were to be summoned to the closing scene, and in truth, those who love him best ought not to wish his sufferings to be prolonged,

The case is somewhat different with our dear Mrs. Kingsley : though she is said by the doctors to be dying, and she may pass away sooner than Henry,—though it is sad and grievous to think that *we* shall not see her again,—yet there is something —comparatively—almost soothing in the thought of her easy and painless descent towards the grave. But the desolation which her death will bring on her poor husband and children is most melancholy to contemplate.

Charles Lyell was (all things considered) remarkably well before this unfortunate fall, which was more alarming than we were at first aware of, and for the time seriously endangered his life. Happily he seems to be in a fair way to recovery, though not yet well, nor (I believe), past the necessity of continued precaution.

In the course of this year we have lost two good old friends in a more humble station :—Mrs. Stocks and Richard Palmer, but in both cases their departure had been long anticipated.

A former friend, but one whom I had not seen for

a great many years (and whom Fanny had never 1874. seen), has departed:—Lady Northesk, formerly Georgina Elliot.

Acquaintances whom we have lost in the course of the year, were: — Herman Merivale, Lady Houghton, John Phillips, Mrs. Hooker, Colonel Yorke, Mr. Charles Austin, Mr. Twistleton and Lady Rich.

The list of eminent persons who have died this year is small compared with that of last year.

Mr. Guizot, M. Van de Weyer, John Phillips (the geologist), Sir W. Jardine, Miss Strickland, are pretty nearly all that I can make out.

On the other hand, the marriage of Sarah Hervey, which promises as well as possible, has been a very great source of delight to me,—I had so long been anxious that she might find a husband worthy of her: and Sarah Seymour's safe passage through the perils of child-birth has been another very happy event.

In the early part of the year Charles Hoare was dangerously ill, and we sympathized deeply with the sorrows and anxiety of his sweet wife, but happily the blow has been averted, and he seems to be now in perfect health.

The marriages of Leonard Lyell, of Cecilia Ellice and of Lady Wilhelmina Legge have been other pleasant incidents in which we could cheerfully sympathize with the parties concerned.

I repeat, that I am deeply thankful, as I have great reason to be, for all God's mercies,—for my wife's and my own good health, and for the very

1874. many blessings which we enjoy, especially for the friendship of so many charming and wise and excellent people. As I wrote last year, I think this is an especial occasion for gratitude.

We have enjoyed a great deal of agreeable society this year, and I look back with particular pleasure on our meetings with the Arthur Herveys in May and June, on our intimacy with Lady Lilford and dear Minnie Powys in London, on Susan Horner's stay with us, and on our meeting with the Kingsleys in September.

Several of our parties here have also been very agreeable, as recorded in this journal.

December 30th.

Sarah Sanford has sent us a newspaper (from Wellington), giving a glowing description of the warm and almost triumphal welcome with which she and her husband were received in that town on their return from their wedding tour,—very pleasant to read.

December 31st.

The last accounts of Charles Lyell are not so comfortable : he and his sister Marianne had both caught colds, and were rather ill in consequence,— she, severely. Katharine, however, does not think there is any cause for alarm.

This is a very severe winter. Hard frost and ground covered with snow ever since the 14th, or perhaps I may say since the 11th, and not the least appearance of any abatement of the cold. The 1874. thermometer down to 12 deg. Fahrenheit.

[During this month of December, Sir Charles settled the dilapidation money, owing to the Vicarage of Mildenhall.

Arthur MacMurdo returned from school on the 16th, and three little Campbells, Annie, Finetta and Frank, paid us a visit of four days. And we had a children's party for them on the 17th. We were alone with Arthur on Christmas day,—our servants had a dinner party and much merriment.

Our nephew Herbert arrived on the 31st of this month.

Among those friends who were not mentioned in his journal—who lunched or dined with us, were:— Mrs. Rickards and Mrs. Wilmot, Mrs. Wilson and her daughter Agnes, Captain and Mrs. Horton and the two Miss Brokes, Mrs. and Miss Anderson and our Vicar Mr. Percy Smith and his wife.

We were occupied, while we were alone, arranging the books in the Porch Room, and Sir Charles in arranging the Colorado plants given him by Miss Kingsley.

We received the Communion on the 6th December, and Sir Charles continued reading Kingsley's Westminster Sermons.—F. J. B.]

1875.

JOURNAL.

1875. The new year opens in the midst of an uncommonly severe winter, much the hardest we have had since '61. The thermometer in the garden last night, was, I am told, as low as 4 degrees Faht. Many of our trees and shrubs will suffer.

A pretty good report of Charles Lyell and his sister to-day. Fanny has had a very pleasant, clever, and lively letter from Sarah Seymour; her baby she reports, is very flourishing and vigorous.

A sudden change of weather: the wind changed last night to S.W., there was a thaw through the night, with some heavy rain, and this morning much of the snow is gone. The trees have entirely lost the drapery of frozen snow, in which for some time past they have looked so brilliant and fairy-like.

Fanny has had a very handsome coloured glass window,* the work of Heaton & Butler, put up in our Church at the east end: it represents Our Saviour blessing little children.

---

* The Barton Church is dedicated to the Holy Innocents. This window is dedicated to the memory of two infant sisters of Sir Charles Bunbury and all holy children of two years old and under.—(F J. B.)

Walk with Fanny. We visited the Boys' School. Resumed examinations of Rose Kingsley's dried plants.

Began to examine Sir Thomas Hanmer's MS. on Gardening.

———

I seem to have omitted to mention my nephew Herbert, who freshly returned from India, came to us on the 29th of December and went away again on the 31st. He is in the artillery, and has been two years away, first in British Burmah and afterwards at Secunderabad, near Hyderabad, in the Deccan. He is a very fine young man, and I was very much pleased with his intelligence and the clearness and good sense of his observations on Indian matters.

———

Had a good walk with Fanny, Arthur, and the dog Grabby. Clement arrived.

The news from Eversley is most sad and distressing. Both our dear friends, Charles Kingsley, as well as his wife, are very dangerously ill. He was attacked some days ago with inflammation or congestion of the lungs, hæmorrhage set in, and he is in extreme danger. Henry writes that, as he was informed by the medical man who is attending *him*, Kingsley's case is hopeless; his life hangs on a thread. It is terrible to think of the danger of losing such a man; the loss, not only

1875. to his children, but to us, to his very numerous
personal friends, and to the whole nation, would be
irreparable. May God avert such a calamity! The
situation of those poor girls, threatened with the
loss of both parents at once, makes one's heart
ache to think of.

———

January 13th.

We have some cause for uneasiness about Charles
Lyell, who (Katharine reports to-day) is very weak.
For some time after his original fall, he seemed to
be recovering rapidly, but soon after Christmas Day
(I think) he caught cold, which seems to have
thrown him back very much, and brought on this
uncomfortable weakness. There is however no
reason to despond. To-day's report of the dear
Kingsleys (from Mrs. Martineau) is rather worse
than better. I fear there is scarcely a ray of hope.
Mrs. Byrne, Fanny's aunt (86 years old) has been
ill, from catching cold on New Year's eve, but is
recovering. Clement went away. Dear Minnie
Napier arrived ; she brings a favourable account of
Sarah and her child.

———

January 16th.

Lady Mary and the Miss Egertons and the
Charles Hoares arrived.

———

Sunday, January 17th.

We went with the Egertons and Charles Hoare
to morning Church. A good sermon from Mr.
Smith

Leopold and Lady Mary Powys, Minnie Powys, Miss Kinloch and her brother, John and Dudley Hervey and Richard Strutt arrived.

Signed agreement with Mr. Robeson, the vicar of Mildenhall, by which he is to occupy my house there (the Manor House), at a nominal rent to the end of this year.

———

Tuesday, 19th.

Had a pleasant walk with Minnie, Mary Powys and Miss Kinloch.

Fanny and most of the party went to the Bury Ball—the two Minnies and I stayed at home.

———

Wednesday, 20th.

The Charles Hoares and the two Lady Marys went to lunch at Ickworth.

Again a bad account of Mrs. Byrne:—in consequence, Fanny did not go to Mrs. Horton's Ball, to which all our guests went, including Arthur.

———

January 21st.

Fanny has had a letter from Mrs. Martineau, giving an account of Sir William Gull's* visit to Eversley: she says, his opinion was, that Mr. Kingsley may, very possibly will, recover,—that Mrs. Kingsley may, perhaps, live for some months, but her malady is mortal and recovery hopeless.

* He was sent by the Prince of Wales.

1875. The Egertons and the rest of our party went away. Dear Arthur went back to school.

M. Madame and Madlle. De Bunsen arrived.

Dudley Hervey gave me some valuable dried plants from Java.

Mr. and Lady Frances Pettiward came to luncheon.

All our company went away except the two dear Minnies.

News this morning of a new trouble and anxiety ; Harry Lyell has had a paralytic attack—slight in degree, but belonging (the doctor says) to a bad species. It is probable he will soon get over it, but meanwhile it is a sad aggravation of the troubles of poor Katharine and of his sisters.

A good report of Charles Lyell, and tolerable of Harry.

We had very pleasant company in the house last week :—a company which we had invited before such dark clouds of sorrow and anxiety gathered over, and in spite of those clouds we still enjoyed very much the society of our friends. For real friends, and not mere acquaintances, many of them are. Kate and Charles Hoare, Lady Mary

Egerton and two of her daughters, Leopold and Lady
Mary Powys, Minnie Powys and our dear Minnie
Napier, Miss Kinloch and her brother, John Hervey,
Dudley Hervey, William Hoare; and from Thursday
to Saturday Baron Ernest de Bunsen and Madme.
and Madlle. de Bunsen, — these made up the
society.

Lady Mary Egerton being warmly attached to
the Kingsleys, sympathized thoroughly with our
feelings about them.

My previous favourable impression of Dudley
Hervey is quite confirmed, and I have taken a great
fancy to him : he is a remarkably intelligent and
pleasing young man.

————

January 26th.

Yesterday the sad news came of dear Charles
Kingsley's death : Mrs. Martineau kindly wrote it
to us in a letter which reached us in the morning,
before we saw the *Times*. He died on Saturday,
the 23rd.—It is a terrible and grievous loss.   One
of the finest and noblest spirits I have ever known
has passed away from this earth, and I have lost an
invaluable friend—one that can never be replaced,
For many years past I have truly loved and revered
him ; I delighted in his society, and I scarcely ever
was in his company without learning something
from him.   Much as I like and admire his writings
—many of which I return to again and again with
fresh pleasure—his conversation was much more
delightful even than his books.   Indeed, I hardly

R

1875. think I have ever known a man whose conversation
was so charming,—so rich in matter, so various, so
easy and unassuming, so instructive, yet so free
from dogmatism and from any tinge of the
preaching or lecturing tone, sensibility, humour,
wisdom, so happily blended.   Many and long talks
I have enjoyed with him.   Our talks were often on
topics of natural science, in which he delighted, and
of which he had great knowledge.   He repeatedly
said to me that, if circumstances had allowed him
leisure, botany and natural history in general would
have been his favourite studies.   He seemed to
take great pleasure in examining dried plants with
me, and in discussing the questions which occurred
in this employment.   And in this, as in other
subjects, his remarks were eminently *suggestive*, full
of ideas, constantly starting something which led
one into trains of thought worth following out.   I
shall ever feel grateful for having been allowed to
enjoy the friendship of such a man, and shall
preserve, as especial treasures in my memory, the
times when I was in his company.   It is a comfort
that I cannot charge myself with having neglected
or undervalued my opportunities of enjoying his
society.   He was repeatedly staying with us here,
yet not so often as we wished.   I look back with
particular pleasure to his visit to us in October, '73,
when he appeared to be in good health and
excellent spirits, and to derive much enjoyment
from the society he met here.   The first ten days
of April, 1861, are also very bright in my memory,
and also the time that he and Mrs. Kingsley spent

with us in September, '65. But, indeed, he was always delightful, whether in his gay or grave moods :—" His gravity was sense, his mirth was wit,"—or rather, indeed, *humour*, for that I should say, was one of his characteristic qualities. His way of telling a story, especially the old popular legends of Devonshire and Cornwall, was capital. One of Kingsley's remarkable and charming qualities was the extensiveness of his sympathy, the power he had of finding something congenial to himself in the most various and different sorts of men, and of thus drawing them to himself by an indefinable attraction. He seemed capable of feeling sympathy with the tastes and pursuits of all who were not positively bad. In the best sense he was capable of being " all things to all men."

I have been tempted egotistically to dwell on the qualities which made his company so peculiarly charming to me : indeed, it is hardly necessary to say much of his higher qualities, which are more generally known through his works, but I can safely say that he was one of the best men I have ever known. His conversation was not only delightful, but always had a tendency to make one wiser and better. His influence was, I believe and hope, very extensive, and I am sure it was an influence entirely for good. He is most generally known, I believe, by his novels, but many of his sermons were admirable, especially those in his last volume (the " Westminster Sermons,") and many of those in the two volumes, entitled " Discipline" and the "Water of Life." Many of his Essays, too, are admirable.

1875. I do not know that his extraordinary power of painting scenery, in words, is anywhere better shown than in " My Winter Garden," " Chalk Stream Studies," " From Ocean to Sea," and the description of Clovelly in " North Devon." The biographical sketches of Buchanan, Rondelet and Vesalius in the volume of " Health and Education " are very interesting. In his West Indian book there are many brilliant descriptions and a great quantity of valuable matter relating to natural history, combining, in the manner characteristic of him, poetical force of perception with scientific accuracy. I do not mean that I think Kingsley infallible.—To say nothing of some extravagances in " Alton Locke" and " Yeast," I think him mistaken and even inconsistent with himself in the view he took of the Civil War in America, and seriously mistaken about Governor Eyre. But in both cases supposing he *was* mistaken, I believe he was misled by a chivalrous feeling.

————

January 28th.

There is a very excellent article on Kingsley in the *Guardian* of this week—excellent in spirit and tone and in expression.

Very glad to hear this morning, from Scott, that his son Robert has obtained the situation he sought, of clerk in the National Provincial Park. Scott is so excellent a man and so true a friend to us, that I am heartily glad of any good fortune which befalls him.

A visit from Lady Cullum, Lady Hoste and Mrs.
Burrough.

———

Fanny and Minnie went over to Mildenhall and
returned at 5.

Went on with arrangement of dried plants
from Java.

———

A very fine day. Had a pleasant walk with Minnie
and Fanny and two of the dogs. Went on with my
notes on Sir Thomas Hanmer's MS. and with
arrangement of Dudley Hervey's dried plants from
Java.

Mr. Percy Smith, who had gone to Eversley
to the funeral, drank tea with us yesterday and told
us much that was interesting about that solemnity;
and this morning Fanny had a very interesting
letter from Mrs. Martineau on the same subject.
The funeral must have been a most impressive and
touching scene:—a vast assemblage of people (un-
invited), of all classes and conditions. The service
was read by Dean Stanley, "as none but he could
read it," as Mrs. Martineau says.*

Mrs. Kingsley was at the open window of the
house, near enough to hear the service round the

---

* Mr. Smith tells me that he has repeatedly heard Kingsley express his
horror of an inert and feeble old age, and his wish that he might die in the
full vigour of his faculties.

1875. grave (so closely does the rectory border on the
churchyard) ; she bore up well at the time, Mrs.
Martineau says, but has since been very ill. Dear
Rose and Mary showed extraordinary firmness and
self-command, Mr. Smith tells us ; but Rose looked
very much worn, and no wonder. They charged
him with a loving message to us, "from beside
the grave." May God help them, poor dear girls.

Received a most agreeable letter from Sarah
(Hervey) Sanford,—delightful in its tone of unaffected happiness. It is really cheering at this
time to find that any of one's friends are well and
happy. She quite sympathizes with us about the
Kingsleys, and writes very well about him.

She writes, that old Canon Beadon of Wells, who
who is 97 (or in his 97th year), has had an attack of
bronchitis, but has rallied from it most surprisingly,
and his friends really think that he will get over
it.

My 66th birthday. I feel deeply grateful to the
Almighty for His mercies, and all the blessings I am
permitted to enjoy.

Our servants enjoyed a dance and merry-making,
which were well kept up.

Had a pleasant walk with Fanny and Mimie.

LETTER.

My Dear Katharine,

Very many thanks for your kindness in 1875.
writing to me when you had so much on your mind
and thoughts ; your kind expressions and good
wishes are very gratifying to me. I am grieved that
you can give no better account of dear Harry;
I assure you I do feel very much for all your
sorrows and anxieties, both about him and about
dear Charles Lyell. It has been a most trying time
these last two months; I never remember a time
when so many of our near relations and dearest
friends were seriously ill at once. I do hope and
pray that both Charles and Harry may recover
completely, and that you, dear Katharine, may not
suffer materially in health from all this anxiety and
distress. It has been a sad accumulation and complication of troubles for you. For myself, I feel
deeply how great reason I have to be thankful to
God for His mercies, and first of all for Fanny's and
my own good health (hitherto) through this trying
winter. Not less reason have I to be grateful for so
many good and kind friends, though we have
lately suffered one grievous loss. Kingsley's death
is a great blow; he was not only a man of genius
and of vast knowledge, but a most warm-hearted
and most loveable man, and a true and warm friend
to us. I have seldom known a man whose con-

1875. versation was more delightful; it was so various—
such a happy blending of sensibility, humour, and
wisdom.   There is an excellent though short notice
of him in this month's *Macmillan*, by Sir Arthur
Helps.   It is difficult to feel quite hopeless about
Mrs. Kingsley, in spite of Sir W. Gull's opinion.

Sarah (Hervey) Sanford, who corresponds with
the Miss Osbornes, Mrs Kingsley's nieces, writes to
us that she writes long letters to her sister, Lady
Sidney Osborne — letters which remind them of
former days.   Rose, whose health seemed to be
giving way, has been carried off by Mrs. Max
Müller to Oxford for three or four days, and already
is said to be the better for the change.

Sarah Sanford's letter is in a delightfully happy
tone, quite cheering to read.   We have excellent
accounts also of *the other* Sarah, through her mother,
whose presence here is a great comfort to us both.
I read to Minnie and Fanny in the evenings; we
have gone in this way through Dorothy Wordsworth,
and Lord Holland's Memoirs of the Whig party,
and we have now begun Dean Hook's Lives of the
Archbishops.   I never find the day long enough for
what I want to do.

I hope you have good accounts of Leonard and
Mary and of Frank.   I will not enter upon botany,
as of course you can have neither time nor thoughts
to spare for it, further than just to tell you, that I have
lately had an interesting addition to my herbarium;
—Dudley Hervey gave me a number of plants,
which he had collected on the summit of Pangerango
mountain in Java, 10,000 feet above the sea.   It is

the same mountain which is described in Mr. 1875.
Wallace's book.

Good bye, dear Katharine, I earnestly hope your
anxiety about Harry will soon be relieved.

Ever your affectionate brother,

CHARLES J. F. BUNBURY.

JOURNAL.

February 6th.

We received the sad news of the death of Harry
Lyell ;—unexpected, for though we knew that he
had been attacked by inflammation of the lungs,
and was seriously ill, we had not been led to
apprehend any pressing danger. But yesterday
morning a change for the worse took place, and he
died in the afternoon, quite calmly and peacefully.
He was a good man.* I am grieved for Katharine,
whose health is not strong, and who has had much
anxiety and distress this winter.

LETTER.

Barton, Bury St. Edmund's,
February 6th, '75.

My Dearest Cissy,

Very very many thanks for your letter, so
full of kindness and affection, written on the 3rd.
Your kind expressions are very dear and precious

---

* He was thoroughly a gentleman, and remarkably courteous in his manners
and demeanour towards people of all conditions.

1875. to me—not that I ever feel any doubt of your love
and kindness, but it is always very pleasant to
receive the reiterated assurances of them. I feel
deeply how great reason I have to be thankful to
God—especially for such kind and loving friends as
I have—as well as for Fanny's and my own good
health (so far), through this trying winter.

I am sure you will be shocked, as we were, by the
news we received this morning of the death of
Harry Lyell. He had, as you are aware,
a slight paralytic seizure just a fortnight ago, having
been apparently quite well before, he seemed to be
recovering from it, though slowly ; a very few days
ago he was attacked by inflammation of the lungs :
yesterday morning the disease took a bad turn, and
in the afternoon he died : I am very much grieved
for poor dear Katharine, who is very delicate, and
who has all this winter been harassed by so many
anxieties, and I am rather afraid of the effect this
shock may have on Charles Lyell, whose state is
precarious.   It is altogether very sad.

Mrs. Byrne is still alive, but generally unconscious ; life seems to be slowly ebbing away, and
under such circumstances one cannot wish mere
bodily existence to be protracted.

Pray thank Henry for his interesting, though sad
letter of the 2nd.   I trust he has quite recovered
from the effects of the fall.

I think you are very right in wishing Emmie to
be amused and to have cheerful society, and I am
sure Henry wishes it too.   I shall be very glad when
I hear of Harry's arrival in England.

Fanny is tolerably well on the whole, though of 1875.
course a good deal shaken by the news to-day.
It is a great comfort that dear Minnie is with us.

Ever your very affectionate brother,

CHARLES J. F. BUNBURY.

JOURNAL.

February 12th.

Mrs. Martineau writes: — "I have seen Mrs.
Kingsley to-day for the first time for more than a
month. She has rallied wonderfully, far more than
was ever anticipated, and looks far more like herself
than when I saw her last." But Mrs. Martineau
warns us not to suppose that this "rally" means
"recovery," "the disease is still there," and this is
only a respite. She may, however, live some
months.

Mr. Bowyer arrived and I walked about the
garden with him.

February 13th.

Mr. Bowyer engaged at the Workhouse at Bury.

February 14th.

Read to Fanny and Minnie Dean Stanley's
Sermon in memory of Kingsley, preached in
Westminster Abbey on January the 31st.—It is
most beautiful: an exquisite composition and a
most true and just appreciation of the man.

LETTER.

My Dear Henry,

1875.   You may well imagine how much we were shocked and grieved this morning by the news of the sudden *death of Lady Cullum.* It was brought to us by Dr. Macnab, who had been in attendance on her. She had seemed very well for some time past, seemed to have quite recovered from her illness of last year, and to be quite like her former self, so that we hoped we should have her among us for a long time. Yesterday, it is said, she appeared to be particularly well, and in the highest spirits, and walked a great deal; Fish and Barrett told Dr. Macnab that they thought they had never seen her looking better, or more lively and animated. The attack must have come on while she was resting after her walk. Dr. Macnab being called in, found her insensible, and she continued unconscious, in a state of *coma* till this morning, when she died. A blood-vessel had burst on the brain.

It is a great blow, a grievous loss, to us and to all the neighbourhood. I hardly know anyone whose loss would be more generally or widely felt throughout this part of the country. She was a delightful neighbour, so genial, so thoroughly good natured and warm hearted, so full of kindness and true goodness of heart; so amusing and full of

fun, with the most perfect good nature at the 1875.
same time ; never betrayed by her lively spirits
and drollery into saying anything that could hurt
one's feelings : so much humour and good humour
both ; her loss to this neighbourhood can never
be replaced.

What a multitude of deaths and illnesses all
around us, this winter. I never remember anything like it. Dean Stanley touched very impressively on this subject in his beautiful—very beautiful
—sermon on Kingsley, in Westminster Abbey.

Fanny had intended to go up to Town this week
to see Katharine, but she caught cold on Saturday
last, and has been quite laid up since with a severe
sore throat, a thing not to be trifled with. She is
better, but will not be able to go out for several
days yet.

Dear Minnie is an incalculable comfort to us, and
she has very good accounts from Sarah.

With my best love to dear Cissy and Emmie,

   Believe me ever,

     Your very affectionate brother,

      CHARLES J. F. BUNBURY.

JOURNAL.

February 19th.

Extremely cold—snow falling much of the day.

My Barton Rent Audit — very satisfactory :
luncheon afterwards with the tenants and Mr.
Smith.

LETTER.

My Dear Katharine,

1875.   What a sad and melancholy account it is of dear Charles Lyell, in your note and Miss Lyell's, just received. I very very much fear that we shall never see him again—not that your accounts appear absolutely hopeless, but the hope seems to be very slight. It is very grievous—very grievous indeed. I very much fear that I shall soon have to mourn the loss of one of the best and truest friends that any man ever had. I feel deeply for you and for his sisters, and for Leonard who has been like a son to him. What an accumulation of sorrows you have had in the last two months! I little thought when I last wrote to you, dear Katharine, that the end of your dear husband's life was so near. I have not written to you since, but I trust you know me too well to suppose that I have been indifferent or insensible to your sorrows. But I knew that Fanny could speak for both of us, better than I could for myself. So good a man and so thorough a gentleman, and one so kind and courteous to all, must certainly be much lamented by all who knew him. Your children will be your great comforts, especially dear Rosamond, and Leonard, who is a young man that any mother might well be proud of, and his wife, too, I am sure, will be all that you could wish.

I am sorry to say that Fanny is very ill, with 1875. bronchial catarrh, Dr. Macnab calls it : he assures me that there is no danger at present, but says that the great object is to prevent its passing into actual *bronchitis*, which is always serious. She is to remain in bed all to-day, and Mrs. Wallis (the housekeeper) is to sleep in her room to keep up the warmth constantly, and that she may not have to get up to make up the fire. Her cough is very distressing at night. It is the same cough which she had in London in '66 and '71. Dr. Macnab gives me hopes that (with strict prudence), she will be much better in two or three days, but she requires great care, and it must be a long time before she will be able to go out.

There is another sad thing : poor Mr. Flood, Lady Cullum's brother, was recalled suddenly from Hardwick to Ireland by the telegraphic news of his wife's dangerous illness, and I am afraid the last reports of her are very bad. It will be too sad if he loses his sister and his wife both at once. What a fatal winter to so many !

Ever your very affectionate brother,
CHARLES J. F. BUNBURY.

P.S.—Fanny has been reading my letter, and begs you will not be anxious about her. She thinks I have made too much of her illness. It is a comfort that the Doctor says her pulse is very good.

JOURNAL.

1875. Nothing to record but deaths and sorrows. Charles Lyell died on the 23rd. This was not unexpected; the report we had had of him the day before had been as bad nearly as possible; and for some time past the accounts of him had been such as quite to prepare us for the end. In the state he had fallen into, we could hardly wish it otherwise. It would have been much more deplorable if he had lingered long in a helpless half living state, with his noble mind weakened and clouded. Death was to him a happy release. Katharine writes that the doctors believe his accident (falling down stairs) on the 9th of December, to have been the real though remote cause of his death. He certainly never has been well since, and they think that there has on the whole been a progressive decline of vital power, though once or twice he appeared to rally remarkably for a time.

I have so often spoken of Charles Lyell in my journal, so often recorded my admiration and affection for him, so often noted his remarks and opinions that I need hardly now dwell at length on his merits. Ever since I married, he has been a truly invaluable friend to me and to my wife; no man ever could have a better friend than I have lost in him. He was by no means fond of business, nor willing to be distracted from his scientific pursuits; but he would make any sacrifice to serve a friend.

Of late years he was much broken, and though his mind continued clear, and his interest in science never abated, his memory sometimes failed, his powers of utterance were enfeebled, and the infirmity of his eyesight (rather than of his limbs) hindered him from any active exertion. But I love to think of him in his best days—to dwell on the memory of what he was when we were so much together at Mildenhall, at the Geological Society in Madeira and the Canaries, and later, when he was newly returned from his adventurous exploration of Etna. At that time (November, 1858) after much conversation on geology and other branches of science, I noted down on his 61st birthday, "that it is a great satisfaction to see him in such health and spirits, so vigorous and active in body and mind." In those days he had amazing energy and activity, both of body and mind, and shrank from no fatigue and no danger in the pursuit of his favourite science. The imperfection of his sight (for he was always extremely short-sighted since I have known him, and not strong-sighted), increased the danger of his geological clamberings. I remember how his companions and the people of the country were astonished, in Madeira, at the way in which he would stand on the edge of a precipice, with his glass to his eye, expounding the structure of the rocks in a sort of extempore lecture, seemingly perfectly forgetful that a slight movement would send him sheer down several hundred feet. His expedition to Palma in 1854, and that to Etna in '58, involved danger enough to make his dear wife very uneasy—

1875. for she was unable to accompany him; but she never shrank from any sacrifice which he wished for.

It is unnecessary to say that Charles Lyell was devoted to the pursuit of geology; he was devoted to it with an ardent and steady passion like that of Linnæus for natural history, or of Nelson for the naval profession. But he took a lively and enlightened interest in botany also, and in every branch of natural science; and he showed the same philosophical spirit in his views respecting all of them. It is very true what Joseph Hooker said of him, that he was the most philosophical of geologists and one of the best of men.

His Liberalism never approached the borders of socialism or communism, nor did it prevent him from being heartily loyal to the Queen and pleased with attentions from her. Prince Albert sought his society and appreciated his genius; and Charles Lyell, long before the Prince was generally understood or valued in England, always thought very highly of him.

The day of Charles Lyell's funeral.

LETTERS.

Barton,
February 27th, '75.

My Dear Katharine,

Thank you much for your kind letter, with the particulars about the arrangements for dear Charles Lyell's funeral. If we had been in town

and the weather had been tolerably mild, I should 1875.
have been very glad to show my respect for Charles'
memory, by attending the mournfully interesting
ceremony. But I do not think it would be right to
add to Fanny's troubles and anxieties, by making
myself ill, which I certainly should be. It is very
unfortunate that the day has turned out so terribly
cold. I am afraid many will suffer from it. I am
astonished to hear that Sir Edward Ryan meant to
take part in the procession — at his age it seems a
desperate risk.

Your friend Lady Smith is indeed a wonderful
woman ; if she lives through this winter, she will
surely be immortal.

(February 28th.) I have been deeply interested
by what you tell us in your letter to Fanny, about
the funeral. The scene in the Abbey must have
been very interesting, and the situation seems very
well chosen.

Fanny, I am happy to say, seems very decidedly
better.

Dear Minnie Napier's company has been most
cheering and an inestimable comfort to us both, but
I am afraid she must leave us the day after to-
morrow :—this is quite right, as she has business of
her own to attend to, but we shall miss her dreadfully.

Pray take care of yourself dear Katharine, and do
not catch cold or any illness, to which I am sure
you will be the more liable from all these causes
producing depression of spirits. I shall be very
anxious to hear of Leonard and Arthur and all your

1875. friends who have attended the funeral that they have escaped the effects of the cold.

Believe me ever,

Your very affectionate brother,

CHARLES J. F. BUNBURY.

———

Barton,
February 28th, 75.

My Dear Susan,

I have long delayed to answer a very kind letter which I received from you on my birthday, and for which I owe you a great many thanks. But now I write to you, not only to thank you for that letter, but to acknowledge for Fanny one which she had from you this morning. I thank God, she seems to be now much better : she had a very good night, hardly coughed at all, and seems altogether better and going on well.—But she is very much weakened and lowered both in strength and spirits, as is natural, as she has been ill a whole fortnight, and most of the time confined to her room,—for two or three days even to her bed, with so much besides to harass her mind and affect her spirits. Indeed, I believe that these depressing influences have had much to do with her illness. It has been a very great comfort to us having dear Minnie Napier's company : she has been most exceedingly kind in staying on with us all through Fanny's illness, and doing everything for her. It is quite reasonable that she should leave us soon, but we shall miss her terribly.

I am afraid Fanny will not get quite well till the weather becomes mild, and that does not seem a very hopeful prospect. We have not had so bad a winter since '61, and then we were 14 years younger and there was not such a succession of deaths of friends to depress us. The last two months have been a sad and fatal time. Charles Lyell, Harry Lyell, Charles Kingsley, Lady Cullum, all gone. We were quite prepared for Charles Lyell's death by the accounts which had come for some time past ; there seemed no prospect of real recovery, and one could not wish him to linger on, half alive, with his noble mind clouded and obscured. His body was laid in Westminster Abbey yesterday, and it is very right that he should be commemorated there. But my thoughts go back to his bright days, and I love to dwell upon the memory of him as he was at Madeira and Teneriffe, and so often at Mildenhall, and in our first visit to Kinnordy and in later days at the scientific meeting at Bath. What intense activity and energy he had, and what a joy in the exercise both of mind and body, what eagerness, and what an interest in all that interested his friends! There could not be a better friend than he was to me. Kingsley, again, was a friend whom I loved and admired and valued exceedingly, and as he was ten years younger than me, I hoped to enjoy his company many a time in the years that might remain to me. Lyell had done his work, and done it nobly ; but Kingsley if health had not failed him, was capable yet of doing much more.

(The remainder of this letter is lost.)

JOURNAL.

1875. Scott gave me a good report of the Mildenhall rent audit.

Read to Fanny and Minnie, some of Dr. Hook's Archbishops.

———

Charles Lyell's mortal body was deposited in Westminster Abbey on Saturday, the 27th. It is a just tribute to his scientific eminence and his great qualities. The funeral was attended by a number of the most eminent men, and must have been a very interesting scene; but Fanny's illness, and the extreme severity of the weather, prevented us from going up to London for it.

Dear Minnie Napier left us to return to London; she had been with us ever since the 13th January— a most delightful companion, and an inestimable comfort to us both.

———

Dear Katharine and Rosamond arrived. Comfortable talk. Fanny came downstairs in the evening—the first time for a fortnight at least.

———

There are excellent articles on Charles Lyell in

The Saturday Review and *The Guardian*, but the best of all I have seen is in *Nature*. This last, I understand is by Professor Hughes of Cambridge.

March 9th.

I have been very sorry to see in the newspapers that Sir Arthur Helps is dead. I did not know him personally, but I am very fond of his writings— especially " The Friends in Council," and I admire his character, and always hoped that I should one day know him. Fanny took a walk with me in the pleasure grounds.

March 10th.

It is a great blessing and comfort to be able to say, now, that my wife is restored to health. Walked with her in the garden.

Visit from Lady Hoste.

March 11th.

Fine bright day. The parrot which had been Lady Cullum's arrived.

March 12th.

Katharine told me that she has, this winter, had two letters of condolence from Lady Smith, Sir James's widow, who in next May will be 102 years of age. The hand-writing was rather illegible, as Lady Smith's eyesight is failing, but the meaning and expression perfectly clear.

Dear Katharine and Rosamond went away.

Resumed my notes on Sir Thomas Hanmer.

March 15th.

Spent most of the afternoon out-of-doors with Fanny, inspecting and labelling young trees.

Went on with notes on Sir Thomas Hanmer.

March 18th.

Business with Scott on Bishop's farm.

March 19th.

Received an interesting and touching letter from Henry.

March 20th.

Wrote a long letter to Henry.

March 22nd.

Lady Susan Milbank, Mrs. and Miss Wilson, and Mr. Prideaux Brune dined with us. A pleasant little party.

Went on with my notes on Sir Thomas Hanmer's MS. on Gardening.

March 23rd.

Quarter Sessions at Bury—little to do. Gery M. Gibson came to luncheon.

More deaths. We have been shocked and grieved by hearing of the sudden—very sudden—death of Emma Rowley. Sir Charles's eldest daughter, and to my thinking, very interesting. And this same day we read in the newspapers, the death, after a very short illness, of the Comte de Jarnac. I never saw him, but have very often heard of him from Cissy and others of the Napier clan ; and I believe he was a man to be very much regretted.

———

A welcome change of weather—quite mild.

Parish business.—Signed the Churchwardens' accounts. Wrote to Sarah Sanford. Walked with Fanny. Afterwards went out in the pony carriage.

———

Clement went away early.

Arranged some more of Rose Kingsley's dried plants.

Fanny read to me some of her father's Journal of 1816.

———

Good Friday. The weather being milder, we both went to morning Church, the first time Fanny had been to Church since her illness. Afterwards we walked in the grounds with the little dogs..

———

Weather again very cold and stormy, with north wind, and a real snowstorm in the morning. Read prayers with Fanny and read to her Stanley's fine memorial sermon on Charles Lyell.

———

March 31st.

Very cold. We drove out in the pony carriage
accompanied by Boy.*

Mr. and Mrs. Robeson arrived. Mr. Beckford
Bevan and his daughter, and Mr. Tom Thornhill
also dined with us.

April 1st.

Weather milder. Lady Susan Milbank came to
luncheon — the Robesons stayed till late in the
afternoon. Mr. Lott came, and Archdeacon and
Mrs. Chapman also dined with us. The Archdeacon
and Mr. Lott very agreeable.

April 2nd.

Dean Stanley's funeral sermon on Charles Lyell,
preached in Westminster Abbey, the day after his
funeral, and now published in *Good Words*, under
the title " The Religious Aspect of Geology," is
admirable, — not inferior to that on Charles
Kingsley.

April 4th.

We went to morning Church and received the
Sacrament.

April 6th.

We went over to Mildenhall, and assisted at the
ceremony of "laying the first stone" of a new

* A colley dog.

Church at Aspal, between the two hamlets of 1875. Beck and Holywell Rows. The Church is much wanted, the subscription for it is good, and the situation well chosen ; and I hope and believe it will do good. The service performed on the occasion was rather long, but Mr. Robeson read it very well.

We had luncheon with the Robesons before the ceremony, tea after it, and returned home to dinner.

April 7th.

Rain at last. Dear Arthur arrived from school. Read part of Mr. Horner's paper on the Geology of Bon. Looking over Rose Kingsley's Colorado plants.

April 12th.

Susan Horner writes under date of April 8th, that some acquaintance of theirs, attempting to go northward from Florence, had found the streets of Bologna choked with snow, two *metres* depth of snow on the roofs of the houses there, and the plains of Lombardy white like Russia. At Florence there was no snow, but the weather was unusually cold. Here, there has been no snow for several weeks, except a storm on Easter Sunday morning which did not lie ; but the cold winds have been most pertinacious.

April 14th.

Wrote an inscription for Palmer's tombstone.

Very fine but cold wind.

Long consultation about plans for new hothouses.

April 16th.

My nephew Harry arrived.

Mr. James Bevan and his daughter Evelyn (a very nice girl) came to luncheon and spent the afternoon with us.

April 17th.

Arthur attacked with chicken-pox. Dr. Macnab came to see him, and we put off our departure for London.

April 19th.

A beautiful day—quite warm.

Harry went to Cambridge to see Mr. Blore and Mr. Newton and returned to dinner. We lounged with Arthur. Fanny's Thibet goats very diverting.

April 20th.

Shocked and saddened by the news in the papers of the 17th, of the death of Lord Alfred Hervey,— No particulars stated. I wrote to the Bishop of Bath and Wells to express our sorrow, and to-day I have had an interesting and touching reply from him. Lord Alfred's death was very sudden, totally unexpected. He had been attending to his official duties till 3 or 4 o'clock on the day of his death, and was returning with Lady Alfred, in Lady 1875. North's carriage to the villa at Putney, where they were on a visit, when " he was seized with apoplexy, and in a few moments became insensible and remained so till his death five hours after." He was only in his 59th year. I did not know him intimately, much less well than his brothers Arthur and Charles, but all I ever heard of him was in his favour. He had a most engaging countenance and manner—the Hervey charm of manner.

The Bishop writes to me :—"The brother we have just lost so suddenly was indeed most excellent — a thorough Christian in faith and practice, upright, kind and exemplary in all the relations of life."

The Bishop mentions at the end of his letter what I am very sorry for, that his son-in-law Mr. Ayshford Sanford (Sarah's husband), is ill with typhoid fever, though in a mild form. This makes one very anxious.

———— ————

April 22nd.

Up to London to 48, Eaton Place, my nephew Harry Bunbury and Arthur MacMurdo with us.

The weather, after a few very warm, bright and pleasant days has returned to all the bitterness and malignity of winter.

Visits from Leonard Lyell, Charles Hoare and Edward.

Very satisfactory answers from Mr. Patrick and Mr. Girdlestone.*

* On Harry's entering the University of Cambridge.

April 23rd.

Weather gloomy and very cold.

Visits from Susan MacMurdo, George and Mr. Philpott.

We drove to Harley Street and saw Leonard Lyell.

April 24th.

Fine but cold.

Fanny went to luncheon with Kate Hoare. George Bunbury came to breakfast and luncheon with us. Harry went to Marchfield. I had a walk with Sarah Seymour.

Fanny had a letter from Sarah Sanford (in answer to one of sympathy which she had written), giving a very good account of her husband, who is going on as well as is possible in such a malady, with no *complications*, and it is hoped that he will soon be convalescent. But, as she truly says, it is at best a terrible illness, and I do not wonder that she was (as she says) shocked when she first heard what his ailment was.

We dined with the MacMurdos at Rose Bank:— met (besides others), Mr. Rucker, a celebrated cultivator of Orchids. Mr. Rucker had brought for the MacMurdos a basket full of the most magnificent Orchids and other tropical plants and ferns —wonderfully beautiful.

Emily ("Mimi") MacMurdo is become a really beautiful girl.

Edward came in late last night to tell us of a piece of news he had just heard—that Helen Richardson is going to be married to Sir Edward Blackett.

A visit from Montagu MacMurdo—very agreeable as he always is. He told us many anecdotes of Mr. Bidder, formerly the celebrated "calculating boy," now at the head of his profession as a civil engineer and "consulting actuary," I think this was the term used),. to the Life Insurance companies. His mature manhood appears to have quite fulfilled the promise of his boyhood: his rapidity and accuracy in arithmetical calculations seem almost miraculous.

MacMurdos observations on Michael Angelo's military engineering. M. A. mixed a great deal of *flax* with the clay in making bricks for building forts, a very wise plan in MacMurdo's opinion, as the bricks would thus be rendered more tough and elastic, and less easily shattered by cannon shot. The mud walls of Indian forts, in like manner, difficult to breach with artillery, owing to the mixture of straw with the mud.

Monday, 26th.

Fine but cold wind.

Engaged most of the morning in giving directions about the binding of books.

Visits from Isabel Hervey, Mrs. Young, Mr. Hutchings. Called on Lady Mary Egerton.

Visited Mr. Walrond.

Tuesday, 27th.

A really beautiful day, quite like summer. Harry came back from Marchfield.

Isabel Hervey came to luncheon—charming. We visited dear Kate Hoare, charming as ever. Met John Moore at the Athenæum.

Wednesday, 28th.

We four (Harry and Arthur being of the party) went to the Lyceum, to see Irving in Hamlet. I was delighted. Hamlet is one of my most especial favourites among Shakespeare's plays, and I read it very often but I had never before seen it on the stage, and I was exceedingly interested. Perhaps I must own that if Mr. Irving had been rather taller and handsomer, he would have come up more completely to my idea of Hamlet (which is somewhat influenced by the portrait of John Kemble) but his acting was very fine.

In the scene with Ophelia, he softened the harshness of his words by beautiful indications hinting the strength of his love, which (as he allowed us to see) he had difficulty in repressing. In the scene with his mother he was very fine ; and the instructions to the players were admirably well given. Polonius (Mr. Chippendale) was capital ; Ophelia (Miss Isabel Bateman) very pretty, and her madness very graceful and touching.

Lady Head came to luncheon. Visits from Susan MacMurdo and Mimi ; Fanny Mallet, Charles Hoare and Edward.

Beautiful weather. Katharine and Rosamond came to luncheon.

— ———

We went, yesterday, with Susan and Mimi Mac-Murdo and Arthur, to see Mr. Rucker's house and gardens at Wandsworth.

Mr. Rucker is a German merchant, long settled in England, a man of great wealth, and evidently of most elegant and refined tastes ; of gentleman-like manners also. He received us very pleasantly, and we spent a most agreeable afternoon in seeing his treasures. To begin with the house—this is full of beautiful and interesting objects, rich, especially in exquisite water-colour drawings, so that to see these alone, thoroughly, would be good occupation for an afternoon. Then, in the drawing room, is a very large glass case, full of the most beautiful and exquisite humming birds, mounted. He seems to have made a special study of these wonderful little creatures, for beside these mounted specimens, he has a great collection, complete (I think he said) according to Gould's work—stuffed skins laid flat in the drawers of cabinets, and systematically arranged according to Gould's method. In another drawer, skins of the different kinds of Birds of Paradise, wonderfully beautiful. The garden is quite a Paradise, not large, but kept in the most exquisite order, and full of beautiful or curious things. The hot-houses, especially, are quite delightful to a botanist ; so full of rare and beautiful things, and all in the most perfect and

T

1875. flourishing condition. The orchids are what he is
most famous for, and of these we saw a great many
curious and very beautiful, though perhaps the
greater number were not now in flower. I was
struck especially with a Dendrobium of marvellous
beauty—Devonianum, I think, combining variety and
brilliancy of colour with delicacy, beyond almost any
other flower I have seen. Other remarkable orchids;
Phalænopsis in great perfection ; Vanda insignis, a
splendid flower; Odontoglossum Alexandræ, with
long spikes of its lovely flowers, and other (distinct ?)
species, not less beautiful,—Dendrobium nobile, and
several other beauties of the same genus, —Cypripedium caudatum, with its curious, long, pendulous
streamers of perianth ;—Masdevallia, several species
with flowers of curious shapes and delicate colours.

Besides the orchids, many very rare and interesting
plants of other families, tropical or sub-tropical ;
some very fine and remarkable plants of Nepenthes;
Drosera Capensis and some other Droseræ, flourishing in pots filled with Sphagnum kept constantly
moist ;—Sarracenias and Darlingtonia (the former
in flower, the latter not), cultivated in the same
manner as the Droseræ ; — Utricularia montana,
very curious and rare. In cooler houses, a rich
collection of rhododendrons, azaleas and Camellias,
many very beautiful ; the greatest rarity, Rhododendron campylocarpum, with flowers of a very pale
delicate yellow. There is also a fine collection of the
rare and beautiful filmy ferns, Trichomanes,
Hymenophyllum, and Leptopteris, in the most
flourishing condition, especially Trichomanes reni-

forme, most luxuriant. In one of the houses, Mr.
Rucker showed me Adiantum asarifolium, which I
never before saw alive—and indeed, he believes it is
the only living plant of the kind now in Europe.*

May 1st.

We went, Arthur with us, to see dear Henry and
Cissy, at Marchfield. It was a melancholy visit.—
Henry's state is much changed for the worse, since
we saw him in October. It is surprising that he is
still able to walk about, and does not appear very
feeble in his movements. He seems to suffer less
when in the open air, and his chief pleasure is in his
garden. His courage, patience and sweetness of
temper under such terribly prolonged suffering, are
really astonishing.

May 2nd.

Mr. James Gibson Craig came to luncheon.

May 3rd.

Edward dined with us.

My nephew Harry (Henry's eldest son) left us for
Gloucester, where he is to read as a private pupil
with Canon Girdlestone, in preparation for entering
Magdalene College, Cambridge, in October. At his
own earnest request, his father consented to his
leaving the navy, and I have undertaken to direct
his education for the Bar, beginning with Cambridge ; his poor father is entirely disabled by the

---

* We were very sorry to see in the newspaper, in October of this same year,
1875, the death of Mr. Rucker.

1875. miserable state of his own health from attending to any business. As Harry knows very little Latin and no Greek, there was no chance of his being able to prepare for the entrance examination at Trinity by October; I have therefore had his name entered at Magdalene, and it is already on the tutor's books there.

Harry is an exceedingly amiable good-hearted, well disposed lad, gentle, serious and thought-ful, with good abilities (I think) and with a decided turn for mathematical studies ; but indolent.

———

May 4th.

Minnie Powys, May Egerton, Mr. Walrond and Clement dined with us—very pleasant. We visited the National Gallery.

———

May 5th.

Mr. and Mrs. Martineau—dear friends of the Kingsleys, whom I have repeatedly mentioned in this journal, but have hitherto known chiefly by their correspondence, came to luncheon with us.— I like them very much, and hope to cultivate their acquaintance. They told us that Mrs. Kingsley's health seems considerably better than in the winter.

After luncheon the MacMurdos came in from the Drawing Room and many other visitors. Fanny and the MacMurdos, with Arthur, went to the play.

———

Visits from Frank Lyell, Harry Bruce and Mr.
Bentham. A visit from dear Kate Hoare and her
father.

The Louis Mallets, Minnie and Sarah and
Edward dined with us—a very pleasant little party,
but unfortunately Fanny was ill and unable to
appear at dinner. Louis Mallet's conversation
very good. Speaking of Indian affairs, he said,
it is certain that the natives almost throughout
British India look with eager interest to Russia,
and are always on the look-out for news concerning
the proceedings and progress of the Russians.
The danger is not so much of the actual invasion of
British India by a Russian army *(that* he thinks will
hardly happen in our time), but of the agitation
which the advance of the Russians may excite in
the minds of the natives. He said, as he had said
to me before, that the great difficulty and danger of
our Indian Empire lies in the finances.

May 7th.

Fanny still unwell—obliged to give up going with
Arthur to Haileybury ; dear Arthur went with
Richard Scott, passed the examination successfully,
and was admitted at Haileybury College. This is a
great satisfaction to us both.

Dear Minnie came to stay with us.

May 8th.

Minnie took me to the Athenæum and brought

1875. me back. I saw there Sir Edward Ryan and Sir J.
Lacaita.

———

May 9th.

Read prayers with Fanny.
Louis Mallet and young Mr. Prideaux Brune
came to luncheon. Clement dined with us.

————

May 10th.

My first introduction to my dear little godson,
Sarah's child, Charlie (Charles Hugh Napier),
Seymour. He is really a superb baby.

————

May 11th.

Fanny went to Haileybury to see Arthur.
Katharine writes that her friend Lady Smith, Sir
James' widow, this day completes her 102nd year,
and is in good health.

═══

LETTER.

48, Eaton Place, S.W.
May 13th, 75.

My Dear Katharine,
        I thank you heartily for your kindness in
offering to give me Harvey's "Nereis Boreali
Americana," which I thankfully accept. Your
choice is judicious as well as kind, for I shall be
very glad to have the book, and it will have
additional value for me as a memorial of your

thoughtful kindness as well as of dear Charles 1875.
Lyell. I should have written to you sooner, but
your note did not arrive till after we had gone to
Marchfield, and I returned in the evening too much
tired to write,—tired more in spirits than in body.
It was a most melancholy visit ; poor Henry is
decidedly worse than when we saw him ten days
ago, and yet I do not feel certain that his release is
very near. The day was beautiful.

I am happy to say that Fanny does not seem at
all the worse—rather the better—for this expedition,
or for that to Haileybury, and I have some hopes
that the change of air may do her good.

We now think of going to Folkestone to-morrow
(Friday) *morning*, and crossing to Boulogne in the
afternoon :* for, as we *must* (it seems) go to Paris,
it seems best to get it over as soon as possible. I
shall be most heartily glad to get back again.

I hope to go and see you soon after our return.

Ever your affectionate brother,

CHARLES J. F. BUNBURY.

JOURNAL.

May 14th.

We, with dear Minnie Napier, went down to
Folkestone—a very good passage to Boulogne.

---

* To visit my aunt, whose health and affairs required the presence of a
near relation.

May 15th.

A splendid day—very hot.

Minnie and I had a walk on the pier before starting. A very hot, dusty, fatiguing, disagreeable journey from Boulogne to Paris.

———

May 18th.

I had a pleasant drive in the afternoon, seeing Rue de Lille, Luxembourg Gardens, Notre Dame, Tour de St. Jacques, R. de Rivoli.

———

May 20th.

The good news of the birth of Leonard Lyell's son.

———

May 21st.

A very pleasant cheering visit from Leopold and Mary Powys. A short visit with Minnie to the Louvre picture gallery.

———

May 22nd.

Leopold and Mary Powys came to luncheon with us—very pleasant.

———

May 24th.

Spent most of the morning at the Salon, were very well entertained—vast number of pictures, some very striking.

———

May 26th.   1875.

We (three), took a long drive by the Boulevard to the Place de la Bastille, and thence by the Tour St. Jacques and the Rue de Rivoli.

———

May 29th.

Dr. Markheim came to luncheon with us. Had a pleasant walk with Minnie in the Champs Elysees.

———

May 31st.

Mrs. Byrne was sent off to Boulogne under care of Dr. Markheim : we had a telegram of her safe arrival.

═══

LETTER.

48, Eaton Place, S.W.
June 5th, 75.

My Dear Katharine,

    I thank you heartily for your kind gift of Harvey's beautiful book,* which is a welcome addition to my library, and which I shall value for your sake and for Charles Lyell's, as well as for its intrinsic merits.

I am very thankful to be at home, and hope it may be a long time before I go out of England again.

* This book had belonged to Charles Lyell's library.

1875.   Leonard was very kind and pleasant, and a great
help to us on our journey from Boulogne, and I was
very glad to hear that the accounts of Mary and the
baby are good.

>   Believe me ever,
>   Your affectionate brother,
>   CHARLES J. F. BUNBURY.

JOURNAL.

Barton, Bury St. Edmund's,
June 12th.

On May 14th, we, with dear Minnie Napier (who
had kindly consented to accompany us), set off for
Paris. The object of our journey was, that Fanny
might make arrangements for the comfort of her old
aunt Mrs. Byrne. The poor old lady, 86 years of
age, had enjoyed perfect health till the first days of
last January, when she had an attack of paralysis,
and for some days was thought to be in great
danger. She has since recovered to a great degree
her bodily strength, but softening of the brain
having set in, she has ever since been in a state of
childishness or imbecility of mind. The charge of
providing for her comfort lies with Fanny, and
hence our expedition (by no means a pleasure trip),
to Paris. Dear Minnie's companionship, with her
constant cheerfulness and kindness, was an inestimable comfort. Happily we got away from
Paris at last, on the 1st of June, and by Fanny's good
arrangements, aided by a very valuable medical man,

Dr. Markheim, succeeded in bringing Mrs. Byrne safe to London, and establishing her comfortably in lodgings there.

June 7th we went again to Marchfield and spent some hours with Henry and Cissy—Henry rather worse than when we were last there, yet it is wonderful how slowly his strength gives way.

On June 8th I had a visit from Canon Girdlestone and an interesting talk about Harry.

I wrote to Henry. On June 9th we came down here — to Barton, and very happy and thankful I am to find myself here again in the tranquillity and beauty of the country. It is most soothing to us both, and of very great advantage to Fanny's health. The beauty of early summer, the delicate varied colouring of the young leaves are past, but the roses are in their glory, and the gardens altogether full of beauty, and the groves full of the song of birds.

The business which harassed us during our stay at Paris prevented us from seeing or enjoying much of what might have afforded us amusement if we had been of easier mind. Some few things however we did see. We had not been at Paris since '67, and of course had not before seen the traces of the war. The general appearance of things, however, did not appear much changed. Here and there traces of the doings of the Commune are conspicuous:—the ruins of the Tuileries, the empty shell of an immense building (the Cour des Comptes, I believe) in the Rue de Lille and some other marks of destruction in that street, the

1875. extensive gap left by the burning of the Ministère des Finances at the west end of the Rue de Rivoli and some similar memorials. But otherwise the chief alteration I observed, was the ostentatious appearance of the words "*Liberté, Egalité, Fraternité,*" on all the public buildings.

I went twice—once alone, and again with Minnie to the Salon, the annual exhibition of pictures. I was struck, as in '67 (though perhaps rather less forcibly) with the very wide range of merit and the audacity shown (as compared with English exhibitions), both in the choice and the treatment of subjects. There are two extraordinary pictures, which at once attract attention, both by their enormous size and by their character, both supremely horrible in subject and in treatment, and terribly powerful. One is by *G. Becker*—"Rizpah keeping off the Birds of Prey from the Crucified Corpses of the Seven Grandsons of Saul" (see II. Samuel, ch. 21, v. 8-11) ; the other by *Gustave Doré,* " The Circle of Dante's Inferno," in which the guilty are tormented by serpents. Both are hideous and horrible — the Rizpah, I think, the more revolting of the two. Numerous battle-pieces, mostly incidents of the war of 1870-71, many of them very clever and having great appearance of truth. Some very good landscapes, and in particular excellent sea-coast views. Some of the portraits also very good. We could not manage a satisfactory visit to the Louvre. We drove several times in the Bois de Boulogne, which is now very pleasant, the damage done in the war seems to have been surprisingly repaired, and little trace of it is 1875. now apparent. The Bois (apart from the mere fashionable drive), is a real woodland scene, very fresh, pretty and pleasant, the trees now in all the fresh beauty of their spring foliage. There is certainly a large proportion of young trees, but also a good many large ones A great multitude of "acacia" trees (Robinia), at this time in most profuse blossom, the air far and wide absolutely loaded with their perfume. I do not know that I ever elsewhere saw the Robinia flowering so profusely. Nightingales, in abundance, were singing delightfully in the trees, not at all disturbed by the carriages passing underneath.

One day I went with Minnie to the Jardin des Plantes, with an order which admitted us to the hothouses. The houses are better stored, and some of the plants finer than what I expected to see, as all were destroyed in the siege ; but of course under these circumstances, it would not be fair to compare the collection with Kew, or with some others. There are, however, some very fine palms and aroids, an uncommonly good plant of the coffee, and a very fine Quisqualis in full blossom. In the open air, I noticed the extensive and instructive collection of hardy plants, judiciously arranged with a view to study, according to the natural orders—very useful. The famous old cedar, planted by Bernard de Jussieu in 1755, appears to be still in good condition. On the 22nd of May, we dined with Madame de Tourgueneff in the Rue de Lille ; the company (besides the family and us

1875. three) comprised M. and Mdme. de Laugel, M. and Mdme. Mohl, Mdme. Mouravief (whose husband annexed the Amoor territory to Russia), a Russian gentleman whose name I did not catch, who had repeatedly travelled in Persia, and was thought to be peculiarly well acquainted with that country.— I was very fortunate, being placed between Mme. Tourgueneff and Mme. Laugel, both very agreeable.

Another day we visited Mme. Tourgueneff, at her country house, Vertbois, near the village or little town of Bougival about seven miles from Paris.   It is very well situated on the steep side of a hill, part of the same range on which St. Germain' stands overlooking the valley of the Seine.   From the upper part of the gardens of Vertbois, there is an extensive and pleasing view over the rich plain of the Seine, studded with villages and scattered houses.   Arches of the aqueduct of Marly seen on the brow of the hills, and St. Germain, further off along the same brow.

The gardens and rich woods of the famous *Malmaison* lie at the foot of the hill, immediately below Vertbois—the two properties touching—but the house is not visible from above.   It is said that there are very fine trees remaining at *Malmaison* from Josephine's time, especially fine deciduous cypresses.

June 9th.

Down to Barton, arriving at home about 9 p.m.

Barton. We went into Bury in the morning. I to Wilson's committee room to hear about the election—saw Mr. Maitland Wilson and his brother Henry ; afterwards we called on Mrs. Rickards.
Finished Vol. 1, of " Lord Shelburne's Life."

———

June 15th.

Went into Bury and voted (by ballot) for Maitland Wilson. Election a very tame affair.

———

June 16th.

Lady Hoste came to luncheon. Satisfactory result of West Suffolk election.

———

June 22nd.

A beautiful day. Up to London. Visit from dear Minnie in the evening.

———

June 23rd.

48, Eaton Place.
We went to Christ Church, Albany Street, to meet the Leonard Lyells and to attend the christening of their baby boy, to whom I am god-father. The ceremony was very well arranged. Those present besides the father and mother and ourselves, were—Katharine, Arthur and Rosamond Lyell, Mrs. Stirling and Miss Emma Stirling. After the ceremony we went to the house in Harley Street (which we had so long known as Charles

1875. Lyell's), and had luncheon with Leonard and his wife, they were very pleasant and the luncheon very nicely arranged.

--- ---

June 24th.

We returned to London the day before yesterday, having spent a fortnight, minus one day, in delightful, quiet, at Barton. We were there, indeed, in the thick of a contested election: the nomination was on the 10th and the polling on the 15th, yet this hardly disturbed our tranquillity. A Suffolk election under the present law is a very quiet—indeed, tame affair. I went into Bury both those days, and found the most profound quiet : and on the 15th I voted for Maitland Wilson, with as little of bustle and agitation as if the voting had been for the officers of a learned society. I had long wished M. Wilson to be Member for West Suffolk, because I know him to be a good man, an honest man, and a sensible and clear-headed man, of moderate and reasonable views, and I believe him likely to be a good and useful representative of an agricultural county. Therefore I gave him my vote, and my interest such as it is. He came in by a satisfactory majority. Declaration of Poll, June 16th :—

Maitland Wilson	..	..	2780
Easton	..	..	..	1061

	Majority	..	1719

On the 19th I received a delightful letter from Sarah Sanford : her husband is quite restored to health, and her letter is delightful, especially in its tone of thorough happiness. They are staying at a

" cottage " belonging to him on the north coast of 1875. Devon—" Woody Bay," near Lynmouth, and she is in raptures with the beauty of the scenery, the vegetation and the rocks. A letter so full of happiness, from so dear a friend, is a delightful refreshment after all our worries.

On the 17th, in the latter part of the afternoon, there was a terrible thunderstorm, one of the most violent that I have witnessed in this country. Fanny returned from Mildenhall through the very worst of it : but thank God she was not hurt. The rain, of which a good deal fell while we were at Barton, was very welcome, and did much good both to the garden and the farm as we had not yet begun to cut our grass. The lawns looked much greener on the last than on the first day of our stay. The roses and honeysuckles were delicious, and the multitude and liveliness of the birds, their beauty and their songs afforded us great enjoyment. I delight in the woodpeckers, the wood-pigeons, the wagtails, redstarts and flycatchers.

June 26th.

A very fine day with one shower.

Minnie Powys and Minnie Napier came to luncheon—Dudley Hervey afterwards, and they three went with us to Wimbledon, to Lady Frere's garden party :—Sir Bartle absent, Miss Frere very pleasant.

We went to morning service at St. John's
Church (near the Victoria railway station), and
heard a very admirable sermon from Mr. Storrs, on
Peace. He dwelt with great felicity of expression
and manner on the numerous failures in Christian
peace and charity—in national concerns, in sects
and churches (touching particularly on the bitterness of "religious" newspapers), and in private and
family life.

In the afternoon, a very agreeable visit from
Norah Aberdare.

We dined with the Charles Hoares—an extremely
pleasant little party: Lady Mary Egerton and her
two eldest daughters, the Palmer Moorewoods, Mr.
William Hoare, Mr. Tait (son of the Archbishop),
and Mr. Cobbold. Kate as charming as ever; I was
most fortunately placed between her and Charlotte
Egerton, also very pleasant. Before the end of
dinner Lady Arthur arrived, just come from Wells.

Fanny went with Katharine to Woking Cemetery
and returned to dinner.

Visited Lady Lilford—very pleasant: afterwards
went and looked at the Royal Academy. Clement
dined with us.

Lady Arthur Hervey came to luncheon with us—
very pleasant.

We visited General and Mrs. Rumley. Edward
dined with us.

Miss Egerton, yesterday, sent Fanny a letter she
has received from Rose Kingsley, with the most
recent account of the family.

Their removal to a house at Byfleet, near
Weybridge, had been successfully managed: Mrs.
Kingsley bore the move well, was pleased with the
place, and all seemed hopeful : but a few days ago,
Mrs. Kingsley had again a most violent and
alarming attack of her heart disease, and for some
time was in extreme danger. The extreme peril
seemed (if I understand rightly), to have passed
away before Rose wrote, but they do not hope to be
able to remove her to another room for a long time.
Poor dear girls, theirs is indeed a hard lot.

Canon Girdlestone called—had a talk with him
about Harry, and advanced money to him for the
tour.

Our dinner party : the Bishop of Bath and Wells,
Lady Mary Egerton and one of her daughters, the
Morewoods, the MacMurdos, Philip and Pamela
Miles, Isabel Hervey and her brother Dudley,
Minnie, Mr. and Mrs. Edward Goodlake, Leonard
Lyell and Harry Bruce, a very pleasant set. After-
wards several other agreeable people came in :

1875. Lady Arthur Hervey, Lady Lilford (not the Dowager, of course), Mrs. Lynedoch Gardiner, Minnie Powys, Mr. and Mrs. Fazakerly, Mrs. Harvey.

———

July 2nd.

Fanny went with Mimi MacMurdo to Haileybury to see Arthur. I had luncheon with Minnie and Cissy Goodlake.

Harry arrived.

— — — —

July 3rd.

John Freeman arrived. We dined with the Brampton Gurdons — met the Thornhills and others.

Went, yesterday, with Minnie to the Royal Academy Exhibition, and had a good look at it. Miss Thompson's picture, "The 28th Regiment at Quatre Bras" (the regiment in square repulsing the French Cavalry), is exceedingly clever, full of life and spirit, with remarkable variety of character and expression in the faces of the men. I asked Mac-Murdo one day, what he thought of Miss Thompson's picture: he said it was very clever, but that, naturally enough, she had fallen into some errors of detail which to a practical soldier take away from the appearance of reality. In particular, the soldiers look much too neat and spruce, their faces much too clean: in the heat of such an action, what with the smoke, the biting of the cartridges, &c., they would become almost as much begrimed

as chimney sweepers. None of the "historical" 1875.
pictures pleased me more than "The Babylonian
Marriage Market," (an illustration of the odd story
in Herodotus about the damsels put up to auction
in the order of their beauty, and the marriage
portions provided for the *unbeautiful* out of the
money given for the beauties),—this by *E. Long*, a
name new to me. Some remarkable portraits by
Ouless, a new artist (whom the Charles Hoares
particularly mentioned to me): Charles Darwin,
capital; but the most striking is the portrait of Mr.
Pochin, an eminent chemist, who is represented
amidst his apparatus, watching the progress of an
experiment — this exceedingly clever. Of the
landscapes, I think not quite so many attractive or
striking as usual, but *V. Cole's* "Lock Scavaig,
Isle of Skye," is very fine indeed, and in quite
a different style from his usual sunny, wide,
extending South of England views. In this style
also his "Richmond Hill," is excellent. I admired
also a picture by *P. Graham*,—a wild moor, with a
storm sweeping across it.

Another excellent picture is "The Last Muster,"
by H. Herkomer : — old Chelsea Pensioners at
church. Cooke's view of Denderah, on the Nile, is
an exquisitely beautiful little thing, delightful for its
transparent clearness and stillness, and the glow
of the sky and the water.

———

July 5th.

Katharine Lyell told us of the gratifying, almost
triumphal, reception which Leonard and his wife

1875. met with at Kinnordy. All the tenants assembling
to meet them at the railway station and escort them
to their new home ;—arches with flowers and inscriptions ;—the horses taken off, and the carriages
drawn by men to the door,—and so forth. This is
very pleasant to hear of.

———

July 6th.

Received a most cordial and interesting letter
from Rose Kingsley—an unexpected pleasure, as
I had not imagined that in the midst of so many
cares and troubles, she would be able to find time
to write to me. She says "her mother has been
terribly ill, but is much better, and the doctors
think that the storm will blow over for this time."
They are very much pleased with their house
(Byfleet, near Weybridge), with the beauty of the
surrounding country, and with their "kind and
charming neighbours." She and her sister had
walked up St. George's Hill, and "got the finest
view I have ever seen in England, — Windsor
Castle and the dear old Buckinghamshire hills
on the north ; the Hind Head and the Guildford
Downs on the south ; over a foreground of magnificent woods, and a middle distance of rich, flat
country." We dined with the Dowager Lady
Lilford—a small party :—Minnie Powys, Lord and
Lady Lilford, Mr. Newton (of the British Museum),
Mr. Holland (a clergyman whom I have heard
preach at Quebec Chapel), and Mr. Best. Lord and
Lady Lilford are lately returned from a long and

very interesting yacht cruise in the eastern part
of the Mediterranean, in the course of which they
visited in particular, Cyprus, which he wished
to explore with a view to its natural history. In
this respect he was rather disappointed, he told me,
for Cyprus, though it has seldom been visited by
naturalists, did not prove rich in novelties or rarities.
Lord Lilford quite confirmed what I had heard long
ago from Edward Forbes, about the extreme beauty
of Lycia and other parts of the coast of Asia Minor
—the grandeur of the mountains brought into
direct and striking contrast with the rich luxuriance
of the low grounds.

July 7th.

Holland House garden party—fine weather and a
pretty sight. We met the Archbishop and Mrs.
Tait, Sir Bartle Frere, Lady Stradbrook, Minnie
Powys, Lord and Lady Lilford, Mr. Hutchings,
Admiral Yelverton, Lady Arran, Sir Charles and
Lady Ellice and others.

July 8th.

Kate and Charles Hoare, Minnie Napier, and
Minnie Powys, came to luncheon with us,—very
agreeable.

July 10th.

Another melancholy and painful visit to March
field—Henry worse.

July 13th.

We called on Mr. Woolner, and saw his clay
model for the bust of Charles Kingsley, which is to
be placed in Westminster Abbey. At first sight I
was disappointed with the likeness, in particular it
struck me that there was a massiveness about the
features and form of the face, not characteristic of
Kingsley. But the longer we looked, the better we
were pleased.

———

July 17th.

We returned yesterday from a visit of a day and
a half to Lord and Lady Rayleigh at Terling, near
Witham in Essex. Our visit was under unfavourable circumstances, for it rained so incessantly and
excessively, that we hardly saw anything but the
inside of the house.

The party was very pleasant—Clara Paley and
her husband (she was more than usually pleasant),
Miss Alice Balfour, a younger sister of Lady Rayleigh, very lively, clever, accomplished and pleasant ;
Miss Kate Frere, very agreeable ; Mr. Clive and his
wife Lady Katherine, also pleasant people : Mrs.
Taylor, a widow, sister of the Dowager Lady
Rayleigh ; an elderly Miss Balfour, a cousin of our
hostess ; and Lord John Hervey.

Several of the ladies spent much of their time in
drawing, in the hall ; Miss Alice Balfour and Miss
Frere being especially good artists. There was
music also. Mrs. Taylor sang delightfully, and
there were indoor games.

Lord Rayleigh showed us his laboratory, several 1875. rooms thoroughly fitted up for the pursuit of physical science, with abundance of apparatus. He showed us several curious experiments in acoustics, which would have been very interesting if we had had any substratum of knowledge on the subject.

July 18th.

My beloved Fanny's 61st birthday.

We went to the Savoy Chapel, where Mr. White preached a very beautiful sermon on the duty of humanity and kindness to animals.

Called on Mrs. Evans Lombe, and had a long and very pleasant chat with her: she is a warm, cordial, and constant friend.

July 20th.

Our second dinner party : Sir Frederick and Lady Grey, Mr. and Mrs. and Miss Locke King, Minnie Powys, Lady Head, Colonel and Mrs. Lynedoch Gardiner, Mr. and Mrs. Longley, Dr. Hooker and his daughter, Mr. Angerstein, Mr. Walrond and Edward—extremely pleasant. Sir Frederick and Lady Grey stayed here till the middle of the next day.

Dr. Hooker said that the collection of hardy out-door, herbaceous plants, arranged according to the natural system, in the *Jardin des plantes* at Paris, is most valuable—"we have nothing so good at Kew." He spoke of the magnificent liberality of the Indian Government in the encouragement of botany and natural history in general.

I went with Fanny to the Royal Academy, and
saw it well.   We met Minnie Powys and her
brother Edward there.   Sir F. Grey is, I think,
rather strongly inclined to the " Low Church "
party, and more desirous than I am of strong
measures against the Ritualists.

———

Mr. White, the preacher at the Savoy (whom we
heard last Sunday) came to luncheon with us ; also
Minnie Powys and her brother Mr. Edward Powys
and Dudley Hervey—a pleasant little party.   Mr.
White is a very interesting man, and my impression
of him from this first acquaintance is, that he is a
wise man.   He is very zealous for the protection of
animals from cruelty.   He spoke with warm
admiration of Mr. Maurice.

Our third dinner party ; Minnie Powys, the three
MacMurdos, Mrs. Maitland Wilson and Agnes,
Augusta and her son John, Admiral Spencer, Mr. and
Mrs. Bevan, Miss Adeane, Captain Adeane, General
Rumley, William Hoare, Edward.—Pleasant.

Just as we assembled for dinner, about eight
o'clock, there came over such a thick and dark veil
of clouds that we could hardly see one another's
faces : it seemed to threaten a most violent storm,
but the storm did not break here.   It is a very
strange season.   So rainy a one certainly has not
been since 1860.

Kate Hoare, Miss Hoare and Admiral Spencer came to luncheon. We dined with the Angersteins. —met Lord and Lady William Graham, Sir George and Lady Bowen, Mrs. Ford and others.

———

July 24th.

With Fanny to the Zoological Gardens—saw the two great tortoises, one of them enormous, lately arrived from the Seychelles Islands: they are of a dusky blackish colour, the upper shell excessively convex.

According to Dr. Gunther (in *Nature*, July 29th), these two tortoises are male and female, though not of the same "breed." The large one which is the male, had been in the Seychelles about 70 years, he weighs about 800 lbs., the length of the shell (measured in a straight line), 5 feet 5 inches, its circumference 8 feet 1 inch, circumference of foreleg 1 foot 11 inches. The female is much smaller. I cannot clearly make out whether Dr. Gunther consider them as absolutely the same species with the great tortoise of the Galapagos. In the parrot house, the variety of form, size, and above all, colour, in what are evidently of the same family, very striking. Remarked as great varieties, male and female, of the great black Banksian cockatoo from W. Australia, of which Henry long ago brought home stuffed specimens (now in my room at Barton), but which I never till now saw alive. Also a still more rare and very strange bird, a black

1875. Cockatoo from New Guinea, with a very peculiarly formed beak. Exquisite beauty of many small paroquets and lories from Australia. Several kinds of Toucan, brilliant in colouring and grotesque in shape. In a small room adjoining the parrot house, the keeper showed us a great rarity, a " Cock of the Rock," (Pipra rupicola, L.), in its full bright orange plumage, with its curious crest, the first I ever saw alive.

_____

July 26th.

A beautiful day—quite warm, no rain. Another visit to Marchfield, more sad than ever. The decline seems to have become more rapid of late, yet the doctors say that Henry *may* yet live some weeks, possibly even as much as two months, though the end *may* come almost suddenly, any day.

_____

July 27th.

Dear Kate Hoare, Susan and Mimi MacMurdo came to luncheon.

Edward dined with us.

_____

July 28th.

We visited Minnie Powys—very pleasant, as always.

_____

July 29th.

A very warm day.

Harry spent the morning with us, and then 1875.
went back to Marchfield.

Down to Barton, Arthur and Louisa MacMurdo
with us. All well at home—thank God.

———

July 30th.

A fine day, but wind cold in the afternoon.

We went to the Wilsons at Stowlangtoft, to a
garden party and cottage flowershow : met many
acquaintances.

———

July 31st.

Clement arrived.

Read some of Mrs. Fletchers " Autobiography."

Resumed examination of C. Kingsley's dried
plants.

———

Augnst 2nd.

The whole of last week (at least from Monday till
Saturday, including both) was fortunately fine and
dry, which has allowed the haymaking to be concluded successfully. Part of our hay indeed was
destroyed, but when the weather became so
resolutely wet, Scott very wisely stopped the cutting
of the grass in the park, and allowed the remainder
to stand till the weather improved ; then he put on
all the labour we could command, and thus was able
to secured a good deal in fair condition. The end
of the hay harvest thus touches the beginning of the
corn harvest—for Mr. Cooper, in this parish, has
actually begun cutting his wheat.

1875.   Several of the fields of wheat in this neighbourhood are looking very well, but much of the barley I
fear is beaten down and seriously injured.

Thanks to the excessive rains of last month, the
grass looks brilliantly and beautifully green—quite
emerald, such as one usually sees it only in May or
the beginning of June. The trees are in fine foliage.

August 3rd.

A beautiful day.

The Assizes at Bury—I was Foreman of Grand
Jury ; not a heavy calendar. Dined with the Judges
Bramwell and Mellor—agreeable men.

The principal public matter that has engaged
much attention and interest, has been Plimsoll's
vehement outburst in the House of Commons, and
the excitement for and against him in the matter of
unseaworthy ships. I have no doubt that Mr.
Plimsoll has in some respects been rash and hasty,
that he has been over-vehement in the pursuit of his
laudable object, that he has sometimes been wrong,
unjust in his attacks on individuals. But may not
the same be said of all bold reformers of great
abuses ? I conceive it might certainly be said of
the leading anti-slavery agitators.

I do not doubt that Mr. Plimsoll's indignation
has been roused by real and great evils, and that
the agitation which he has set on foot, the attention
which his vehemence has attracted to the subject,
will in time (if not immediately) produce a very.
important reform.

We went to a garden party at Archdeacon Chapman's. Met many whom we knew. Fanny received a most delightful and interesting letter from Sarah Sanford.

Captain Horton came to luncheon and spent the afternoon with us.

Clement went away.

We spent part of the afternoon with Lady Hoste and Mr. Green, and saw their improvements at Nether Hall.

Between Barton and Pakenham I observed the cutting of the wheat begun in several fields.

Mrs. Fane and Miss Spencer came to luncheon. The wheat harvest has begun generally in this parish, as Scott tells me.

We (with Arthur and Louisa), went over to Mildenhall to a garden party at the Robesons—met several acquaintances, returned to dinner.

Between Barton and Mildenhall the wheat harvest is now general, and I understand the crop is thought likely to be quite up to the average.

LETTER.

My Dear Katharine,

1875.   I hope the air and quiet of Forfarshire are doing you good. I have been interested by your notices of Kinnordy in your letters to Fanny. We are enjoying Barton very much, and have had many pleasant days, though as yet nothing like settled summer weather. The catalpa is in glorious blossom, the white lime scents the air far and wide, the new greenhouses are growing up rapidly, and all the pets are flourishing.

I am reading Clements Markham's "Life of Lord Fairfax"—very interesting and well written : also a novel, Mrs. Gaskell's "Wives and Daughters," which I like very much.

Fanny is, I hope, decidedly better than when we first came from London, but she cannot sleep well, and continually complains of fatigue, she has too much to do.

The news from my poor brother is as sad as possible, the only change from bad to worse.

Ever your very affectionate brother,

CHARLES J. F. BUNBURY.

———

JOURNAL.

About the time we returned from France, our

dear Mrs. Kingsley sent me the the whole of her husband's collections of dried plants, including those collected in his youth, as well as in later days in his hasty tours in Wales, Ireland and the South of France. It was a very kind thought of her. They were in great confusion, and the greater part of them damaged by insects, a large proportion, indeed, quite ruined, but I find among them many interesting plants, particularly those which he gathered in the South of France (chiefly near Biarritz), also some from Kerry and Galway, and several rarities (older specimens), from Devonshire and Cornwall. Thus they are welcome additions to my herbarium, independently of their special interest to me, as collected by one whom I esteemed and valued so much, and with whom I have so often conversed on the subject which these specimens serve to illustrate. The most interesting of the specimens are those Biarritz plants, mentioned by Charles Kingsley in his beautiful Essay " From Ocean to Sea :"—Erica ciliaris, Trichonema, Narcissus bulbocodium, Lithospermum prostratum, Smilax aspera, Asphodelus, Daphne cneorum and some others.

August 16th.

Scott tells me that the wheat harvest has begun in Mildenhall Fen, and that the crops look well.

August 17th.

Glorious harvest weather. Talk with Scott.

1875. We had a pleasant little drive in the pony-carriage with the dogs.

Began to revise Brazilian botanical notes. We had a pleasant drive in the barouche. Beautiful evening—splendid clouds.

We drove into Bury and saw Mrs. Rickards.

The wheat on my farm has now all been carried and got in in excellent condition. Scott tells me that Cooper also has finished his wheat harvest most prosperously. The weather all this week, and indeed for nearly a fortnight past, has been superb. In fact, there has hardly been any rain since the beginning of this month.

We drove to Stowlangtoft—saw the Maitland Wilsons and the Byrons and Sir Richard Kindersley.

Fanny went to a garden party at the Phippses at Euston.

I was very sorry to see announced in *The Times* of yesterday, the death of Sir Edward Ryan— sorry, but not very much surprised, for the death of a man in his 82nd year can never be quite unexpected. He was an excellent man, a very 1875. able man, a very valuable public servant, and moreover, a very pleasant man. He was an old friend of Fanny's family, and through them I became acquainted with him.

A visit from the Duke of Grafton and Mr. Coleman.*

Dear Arthur and his sister went away to Rose Bank. We went (by road) to Wherstead Park, to Lord and Lady William Graham's—Colonel and Mrs. Herbert staying there: the Robert Anstruthers dined there.

We returned yesterday afternoon from a three days' visit to Lord and Lady William Graham, at Wherstead, about two miles beyond Ipswich. Wherstead Park is on the high ground on the south-west (or rather, indeed, in that part the west), bank of the Orwell, the highest ground which occupies the space between the Orwell and the Stour, and forms the prettiest and most agreeable part of Suffolk. This Park (Wherstead), is pretty, but by no means comparable to Woolverstone, either for the form of the grounds or the timber. There are, however, many stately old oaks and

---

* Mr. Coleman married a niece of Mr. Edward Ellice.

1875. elms, and some very fine flourishing chesnuts. The house is not remarkable either for size or beauty, (much larger, however, than one would suppose from first sight) — it belonged to the celebrated Admiral Vernon, and there are several relics of him : in particular, a figure of the Virgin, taken out of the chapel of a Spanish ship of the line captured by him ; a model of the "Burford;" bird's-eye views of the harbour of Porto Bello, &c.

Lady William Graham (who was Mrs. Dashwood when our acquaintance with her began), is still very handsome, though her age is over 50,—is very good-natured and good-humoured, cheerful and pleasant. Lord William is an accomplished artist, and draws admirably in water colours, an art to which he is much devoted. We saw a great number of his drawings and coloured sketches of scenery in Italy, the Alps, England and Scotland— very beautiful indeed.

On the 1st of September our host and hostess took us in their open carriage to see Mr. Mills at Stutton, and the next day to the Berners' at Woolverstone. Our dear old friend Mr. Mills (now 84 years old), was looking remarkably well, not at all changed since we last saw him, he talked very cheerfully, seemed to enjoy conversation, and was as full as ever of friendly warmth and cordiality towards us, his mind perfectly clear, and his memory excellent. Mrs. Mills was absent on a visit. Mr. and Mrs. Berners also appeared quite unchanged, and greeted us with all their old frank friendliness.

I have already, in former years, described that 1875. beautiful Park of Woolverstone, and the pleasant house, with its delightful conservatory, which is now in great perfection. The new addition to the conservatory, built within the last two or three years, made in imitation of a natural rocky gorge or chasm in a limestone country, and especially fitted for the cultivation of ferns, is quite delightful to a botanist: the shade, abundant moisture and temperate warmth are specially favourable to the growth of ferns (not absolutely tropical, but those of sub-tropical and warm-temperate climates), which flourish in great vigour and in most picturesque beauty among the rocks.

At the time of our visit, the Bernerses were presiding over a *harvest home*, and a great gathering of their neighbours, rich and poor: there were a variety of games and sports for the men and boys, and abundance of feasting. It was a pleasant sight.

———

September 5th.

We went to morning Church and received the Sacrament.

Very much shocked and grieved, yesterday, by the very unexpected death of Maitland Wilson. He died yesterday (Saturday), afternoon, after a very short illness. He is a great loss to the neighbourhood in many ways. He was an excellent man, most diligent, sensible, and useful as a magistrate, an officer of militia and in all public relations, warmly and justly beloved by all who

1875. were connected with him by private ties. I grieve especially for his wife, who is a charming woman, one of our most agreeable and valuable neighbours, and was quite devoted to him. Indeed, the whole family (including *her* father Sir Richard Kindersley, and her sister and brother-in-law, Mr. and Mrs. Byron), are in a remarkable degree affectionate and united. (Sir Richard and the Byrons are with her now).

September 7th.

A most beautiful day.

Dear Arthur came back from Fulham. Lord Talbot de Malahide and Miss Talbot came to luncheon.

LETTER.

Barton, Bury St. Edmund's.
Sept. 13th, 1875.

My Dear Henry,*

I hope it may perhaps amuse you a little to hear something of the outward condition and appearance of the old place. We have had, in general, beautiful weather since we came back from London—splendid harvest weather, hardly a single shower from the beginning to the end of harvest, and therefore, the crops, such as they were, were got in in the very best condition. As Mr. Greene said the other day, this fine harvest weather

---

* Col. Bunbury died on the 18th September.

has been a great mercy, for the corn had been so 1875. beaten down by the storms of July, that bad weather during harvest would have been perfectly ruinous to the crops. As it is, I believe they are hardly up to the average,—the grains not large nor the ears well filled, owing to the cold and bad weather in the flowering season ; but still we have great reason to be thankful. Scott tells me, too, that the labourers in this parish have behaved particularly well during this harvest time, and the relations between them and the farmers have been more than usually pleasant.

This place was looking most beautifully green when we came down at the end of July, the lawns and park as brilliantly verdant as in May or June, but the dry August has told upon them : the grass is beginning to look much parched, the ponds and ditches are drying up, and the wells beginning to fail. It is in general a very good season for flowers, as was to be expected from so much rain followed by warm and sunny weather, but, I do not know why, our Magnolia grandiflora has scarcely flowered at all. The Catalpa was magnificent, and your Indian horse-chesnut flowered finely in June. The crops of wild fruits are prodigious, crab-apples, haws, acorns, beech-mast. It is not very common for the *purple* beech to fruit, but this year, two of those in the pleasure-ground are bearing *mast* pretty plentifully, just like those of the common sort in size and shape, but richly tinged with purple-brown.

Fanny had a very pleasant letter from Harry this

1875. morning, in the best possible spirit and temper : he must come and stay some time with us here before he goes up to Cambridge. She had also, a day or two ago, a very entertaining letter from Clement, who had just returned to Berlin after an excursion to Dresden and Prague, which he seems to have enjoyed extremely.

Maitland Wilson's almost sudden death was a great shock, very startling to us,—it seemed so strange that he should be cut off so soon after his success at the election. He was an excellent man, and a first-rate man for all county business, and is a great loss to the public, but *we* feel especially for his charming and devoted wife, our most intimate friend in this neighbourhood since Lady Cullum's death. There have been sad gaps of late in the society around us.

Fanny is tolerably, not very well: much engaged at present with Arthur, who is to return to school next Friday. Miss Boileau, a great friend of ours, is here just now. We had expected Annie and Finetta Campbell and two or three of their brothers this evening, but are disappointed.

Of course, dear Henry, you must not think of taking the trouble of writing to me : I have written this merely in hopes of giving you some little amusement by my report of the state of things at Barton. I hope the beautiful weather which has returned to us will enable you to get out-of-doors again and enjoy the beauty of your garden, and I trust you are suffering less than when we were last with you.

With Fanny's earnest love as well as mine, 1875.
believe me ever,

Your truly loving brother,

CHARLES J. F. BUNBURY.

JOURNAL.

September 14th.

Walked about the arboretum with dear Minnie
Boileau, who went away after luncheon.

September 15th.

Continued fine bright weather. Though there
was such an unusually heavy fall of rain in July,
yet we are already again beginning to be parched up
and to suffer inconvenience from scarcity of water;
such is the effect of the almost uninterrupted continuance of fine dry weather from the beginning
of August until now.

Tom Thornhill came to tell us that he had been
urged and had consented to stand for West Suffolk
for the seat vacated by the death of M. Wilson.
There was a meeting of Conservatives at Bury this
morning, and the conclusion was to ask Thornhill to
stand. He is a very good fellow, a thorough gentleman and of some intelligence, but I have never
happened to think of him in the character of a
politician, and I do not know what figure he will
make.

1875.

September 16th.

Captain and Mrs. Ives came to luncheon, and we walked round the arboretum with them.

———

September 17th.

Dear Arthur went back to Haileybury.

———

September 18th.

Splendid weather; very hot and dry. My Barton rent audit, very comfortable and satisfactory. Fanny as well as I met the tenants at luncheon. In the evening a telegram telling us Henry was sinking.

———

September 19th.

We received about 11 a.m., a telegram announcing the death of my dear brother Henry.

———

October 3rd.

Read Kingsley's beautiful sermon, De Profundis.

———

LETTER.

Barton,
October 21st. 1875.

My Dear Katharine,

I have a petition to offer to you on behalf of our dear Mrs. Kingsley. She is making a selection of her husband's letters, with a view to publication ; she thinks that he must have written several letters to Charles Lyell, and she would be very thankful if they could be lent to her (supposing they still exist) that she might have copies taken of them. I sent her the letters that *I* had, and she returned them most carefully and very quickly. I ought to say that she does not mention *your* name— she does not at all know that *you* are arranging Charles Lyell's correspondence—she merely asks me to inquire. There is no hurry at all (this Rose particularly says) ; but if by and bye, you should find that you can with propriety, and without inconvenience, comply with this request, Mrs. Kingsley is one who is worth taking trouble for.

You must have a very interesting occupation, though a laborious one, in revising and arranging Charles's correspondence ; and I am very glad you have undertaken it, and earnestly hope you will have health to carry it through. I trust that your long quiet stay in Scotland has done your health a great deal of good.

Fanny has written to you an account of our visit to Cambridge, and of your Arthur's visit to us here afterwards; and has told you how very favourable an impression we have of him. I hope our Harry will see much of him.

Poor dear Cissy and her daughter are with us now, as well as the MacMurdos and William Napier, (but the last is going away this afternoon). Cissy is wonderfully calm and composed, gentle as she always was, and even able to enjoy a little quiet, cheerful conversation. The weather has been very stormy, wet and changeable, though we have had no frost yet, nothing to cut off even the

1875. tenderest flowers ; and but few of the trees have yet changed colour.

Our new greenhouses or planthouses are getting on rapidly, and are not far from being finished ; those plants from the old houses which were not rooted in the ground are already moved into them. I think they look very well, and I hope you will see them before very long.

Will you tell me whether you would like to have specimens of the following plants ? Rare Cornish or Devonshire plants ; I have put duplicates of them aside for you, out of Kingsley's collections.*—

> Herniaria glabra.
> Corrigiola littoralis.
> Illecebrum verticillatum.
> Sibthorpia Europæa.

There is no hurry about answering this question either ; I will take care of the specimens for you unless I hear that you don't want them.

With my best love to dear Rosamond.

Believe me ever,

Your affectionate brother,

CHARLES J. F. BUNBURY.

JOURNAL.

October 25th.

On the 18th of September, in the evening, we received a telegram that dear Henry was much weaker and sinking, and in the forenoon of Sunday the 19th,

* Mrs. Kingsley had sent Sir Charles Bunbury all Mr. Kingsley's dried plants.

another, telling us that his sufferings were over. 1875.
We arrived at Marchfield on the 21st, and found
there, with dear Cissy, the three boys, Sarah Craig,
William Napier and Edward.

The funeral on the 24th was attended by a great
many persons besides the family,—neighbours and
old military friends. We remained till the 27th.
The house at Marchfield being full, we were lodged
(tolerably comfortably), at the Royal Hotel at
Ascot, from whence we drove over to Marchfield
every morning, returning in the evening.

When the end came, one could not but feel that his
gain so far over-balanced our loss that it ought to be
a subject of thankfulness rather than of sorrow.

The last few days were almost painless, and at
last he passed away so gently and quietly that both
Cissy and Sarah Craig who were watching him
thought he had merely fallen asleep.

Henry's character was a very noble one. I do
not think that a man could have a more genuine
elevation of mind or a higher courage, both personal
and moral, or be more absolutely superior to all low
motives and mean calculations. He had great
practical ability and excellent good sense, his
affections were warm, true and constant. There
could not be a better brother. He had a very
high standard and sense of duty, and though
he was always geutle and kind towards children and
the weak, he might sometimes be stern in his
bearing towards men whose conduct did not come
up to his standard of right. He was, perhaps,
rather too intolerant of stupidity, silliness and

1875. blundering,—certainly, for his own interest, he was too straight-forward and uncompromising in exposing the errors of men in authority. He never would pay court, even by silence to anyone; if he thought that what was said in discussion, or what was proposed to be done, was foolish or mischievous, no matter how high the rank of its author, he would express his opinion plainly and without softening. Hence he was not a favourite with men in office, and his success in his profession was not in proportion to his merits.

On the 27th of September we left the Ascot Hotel, and went (by road not by rail), to the Albert Seymour's house, near Farnborough Station, a very pleasant drive of rather less than two hours, through a very pretty country, by Bagshot. We spent the next day and part of the 29th with Sarah and Albert and Minnie, all of them as cordial and pleasant as possible, but unluckily I was by no means well, and the weather was very bad. The 29th, afternoon, we went by railway from Farnborough to Weybridge Station, drove from thence to the house at Byfleet, where Mrs. Kingsley is now living, passed a delightful hour with Rose (Mrs. Kingsley not well enough to see us), then back to the Weybridge Station, and so by the railway to London.

It was an exceeding great pleasure to see dear Rose again after all the sorrows of this year, and the frank and affectionate warmth of her welcome was very delightful. Her face shows the traces of sorrow and anxiety, yet on the whole I thought her

less altered, and looking better than I had been led 1875.
to apprehend. I was very happy to see her in a
home which seemed so pleasant and comfortable :—
a nice neat house, very prettily fitted up by her
and her mother and sister, with the prints, etc.,
which they had brought from Eversley : a pretty,
quiet, sheltered garden, affording them shade and
verdure and flowers, a greenhouse, the management
of which is a great amusement to Rose, a very
agreeable situation in a very pretty country (at the
foot of St. George's Hill), and as she told me, a
very good neighbourhood.

We remained in Town from the evening of September 29th to October 5th ; during which time
I had the pleasure of seeing dear Kate Hoare twice,
and also her father and mother. Otherwise I did
not enjoy London at this time, for I was not well,
and the weather was very bad.

October 5th, we came down to Barton ; and were
joined on the 7th by Harry, and on the 12th by
dear Cissy and Emily. The 13th we all, except
Cissy, went to Cambridge ; Fanny and I by road,
Emily, Harry, and Clement, by rail ; we visited Mr.
Pattrick, the tutor of Magdalene, a very pleasant-
mannered man ; had a very satisfactory talk with
him, and finally saw Harry establishd in his college
rooms, Fanny settling his furniture for him. I have
since heard, that the rooms, in which Harry is
established, are the very same which Kingsley
occupied while an undergraduate. James Hervey
and Arthur Lyell from Trinity dined with us at the
Bull.

1875.   October 14th, a day of dreadfully bad weather—
of violent rain, wind and cold: in the morning
while Fanny again went shopping for Harry, I went
with Emily and Clement to see the new Chapel of
S. Johns, which I had not seen since it was in an
early stage of building. It is a noble work, grand
and very beautiful, especially the windows, which
are of very deep-coloured glass, and of which the
effect is exceedingly rich, fine and solemn. We
visited also the Chapel and Hall of Trinity, but the
bad weather prevented us from seeing anything of
the College walks. That afternoon, the 14th, we
returned home, leaving Harry at Cambridge.

October 26th.

Susan and Mimi MacMurdo went away.*
Walked with Fanny and Minnie.

October 30th.

Our new plant houses finished, and the plants
from the old houses all arranged in them.

Mrs. Kingsley, in a very beautiful and touching
letter of condolence which she wrote to Fanny some
time ago, had mentioned that she was making a
selection of her husband's letters, with a view to
publication. I told Rose when I saw her at Byfleet,
that I had several letters of his: and finding that
they would be welcome, I sent them to Mrs. Kingsley as soon as we returned home. She had them
copied, and returned them promptly with thanks.

* They had been staying with us for some time.

She also wrote asking me to write down for her 1875. some of my impressions of her husband's character and genius. This I found rather difficult to do to my satisfaction, but I took a good deal of pains about it, working at it for some days after our return from Cambridge, and though what I wrote by no means came up to my own idea, she was much pleased with it, and thanked me very warmly. I never knew a more ardent and thorough-going friend than Mrs. Kingsley.

November 1st.

Mrs. Rickards and Mrs. Wilmot came to luncheon.

Our new plant-houses were finished and brought into use, and the plants moved into them out of the old greenhouses. They look very handsome, and appear (as far as I can yet judge) extremely well-suited to their purpose, and there is ample room in them for additions to the collection. They are in three compartments, in a somewhat cruciform arrangement: the first compartment (nearest to the garden door), intended to be a sort of "orangery," for such plants as require moderate heat, without much moisture, with a free circulation of air,—the middle or *transept* for tropical plants, and the third for ferns in general. They have been constructed by Weekes, of the King's Road. The cost is to be £920, exclusive of the brickwork.

November 3rd.

Meeting at Bury about Prevention of Cruelty to Animals, Mr. Rodwell presiding. Mr. and Mrs. Robeson arrived.

November 7th.

We went to morning Church with Minnie, Sarah and Albert Seymour and the Goodlakes, and received the Sacrament.

Fanny has had a letter—a very pleasant and friendly one—from Louis Mallet, who is about to start for India, on a Government mission connected with the finances. I feel sure that, whatever work he has to do, he will do thoroughly well. His wife goes with him, and I think they are likely to have a very interesting and pleasant excursion. I trust they will both keep their health.

November 8th.

A shooting party in our home plantations.

Lord John Hervey went away in the morning, and Harry after the shooting.

Read Grant Duff in the *Contemporary Review* on India.

November 12th.

Since we returned home, we have had some very pleasant company in a quiet way. Cissy and Emily, William Napier, Montagu, Susan and Mimi and Mr. and Mrs. Edward Goodlake, also our darling Sarah Seymour, and her very amiable and estimable husband, Isabel and Dudley Hervey 1875. (whom I like particularly), were with us from the 3rd to the 6th, Admiral Spencer from the 4th to the 9th. Lord John Hervey, and my nephews Cecil and Herbert, have also been here. Dear Minnie came on the 25th of October, and (with the exception of going for three days to London this week) has been with us ever since.

LETTER.

Barton,
November 14th, 1875.

My dear Katharine,

Fanny is tolerably well, but at present un-happy on account of *Ariel*, one of the lovely little green parrakeets, which is very ill in moulting ; and she fears it will die, poor little thing. We have no company with us just now, except Minnie, and her dear little grandson, Charlie Seymour, who has been left here with his nurse, because there were rumours of scarlet fever at Aldershot, whither his father and mother were obliged to return. He is the most beautiful baby I have ever seen, and very amusing. We celebrated his *first* birthday last Thursday. We have dismally dark and wet weather, with the interlude of a fine day now and then; it reminds me of the November of '52, when Ely returned to its original state of an island, and when Charles Lyell saw from the top of Mildenhall steeple a vast sheet of water taking the place of the Fens.

1875. There is great anxiety and apprehension just now about Mildenhall Fen ; the river is nearly up to the top of the bank, and is rising, and likely to rise, for it rained very hard yesterday afternoon ; and there is great fear that the water will overflow the bank, in which case it would flood a great extent of reclaimed land. Scott is gone there again to-day, to superintend the operations ; but after all that can be done, we are mainly dependent upon the clouds. Here the wet does us no harm.

To-morrow we hope to see the Sanfords—Sarah Hervey and her husband. It will be a very great pleasure ; I have not seen *her* since June of last year, and I am very anxious to make *his* acquaintance. We had a very pleasant party the first ten days of this month.

Poor Cecil is very unhappy (and I am afraid with much reason), about his sweet young wife, who has had a severe relapse.

Fanny sends you her best love.

Believe me ever your affectionate Brother,

CHARLES J. F. BUNBURY.

I have found for you some good specimens of Erica vagans, and one (rather a poor one) of E. ciliaris. When will you come and see our new plant houses ?

———

JOURNAL.

November 15th.

Dear Sarah Sanford and her husband and step-daughter arrived.

November 16th.

We walked with the Sanfords in the morning to show them the gardens, stables and arboretum, in the afternoon Fanny took them into Bury. Most agreeable conversation with them.

November 17th.

Captain and Mrs. Horton, two Miss Brookes and Mr. and Mrs. T. Thornhill dined with us.

November 18th.

A beautiful morning.

A very agreeable stroll with Mr. Bentham, Mr. Sanford and Mr. Wilmot Horton, through the gardens, etc.

Showed part of my collections to Mr. Sanford.

November 19th.

A good report from Mildenhall Fen. Walked with Sarah and Mr. Bentham to show them the dedicated trees. Mr. Bentham went away : all the rest of our party went to luncheon at Livermere.

November 20th.

All our guests, except Minnie, went away—the dear Sanfords last, after luncheon.

November 22nd.

Last week was a delightful one for me, we

1875. enjoyed the company of dear Sarah Sanford and her husband from the Monday to the Saturday.—A most pleasant time it was. I have been very anxious to make acquaintance with Mr. Sanford. He reminds me of Charles Kingsley by the variety of his knowledge and tastes and the wide range of his acquisitions, alike in science, literature, art and antiquities. He has travelled much, and tells his experiences, both of men and of nature in a most interesting way. He was formerly—before his first marriage—Colonial Secretary in Western Australia, where he seems to have observed the aspects of nature and the natural productions of the country very zealously and carefully: I was exceedingly interested by what he told me of the natural history of that strange country. He is a good geologist, and the department to which he has specially devoted himself, and in which he is most deeply learned, is that of fossil Mammalia, and more particularly those of the cavern deposits. He explored, together with Boyd Dawkins, all the caves in Somersetshire, and had a great share in that geologist's work on fossil Mammalia.

One, and not the least important of the many respects in which I find a congeniality and sympathy between Mr. Sanford and myself, is his enthusiastic admiration of Kingsley's writings.

It is delightful to see dear Sarah so happy and united to a man so suitable to her. God grant they may long be preserved to each other.

We had a few other guests at the same time :— Mr. and Mrs. Wilmot Horton and Mr. Bentham.

Mr. Wilmot Horton is a clergyman, was formerly in 1875. Ceylon (of which his father was governor), has seen a great deal, has much various knowledge, and is exceedingly pleasant. George Bentham also was very agreeable, and they both seemed to be very congenial to Mr. Sanford, and he to them.

LETTER.

Barton,
November 22nd, 1875.

My Dear Katharine,

The Sanfords came, and are gone, and the visit to which I have been so much looking forward, is a matter of memory. I have not been at all disappointed; I like Mr. Sanford very much indeed. As far as I can judge from these five days, I do believe that he is really *worthy* to be the husband of Sarah Hervey; and I can say nothing higher of any man. He is certainly one of the most accomplished and interesting men I have met for a long time : a man of a vast variety of knowledge and pursuits— devoted to ornithology, to palæontology (especially the study of fossil mammalia), to painting, to architecture, and antiquities; with a considerable knowledge and love of botany, though it has not been one of his special pursuits. What is still better, he is eminently a gentleman in appearance, in manners, in the tone of his conversation, and (I am sure) in feelings and character. There can be no doubt or mistake about his and his

1875. wife's happiness and devotion to each other. It is delightful to see Sarah so happy, and united to a man so congenial. With them came *his* eldest daughter, *her* step-daughter, Nellie by name, a nice frank, natural, cheerful girl of 17.

Now we have no one with us except dear Minnie and her lovely little grandson, Charlie Seymour. Next week we expect the Bishop and Lady Arthur, and their youngest daughter, and Mr. and Lady Mary Powys. Afterwards, I believe, we shall be quite alone.

We have had and have very bad weather.

With my best love to dear Rosamond,

> Believe me ever,
>> Your affectionate brother,
>>> CHARLES J. F. BUNBURY.

JOURNAL.

November 29th.

Weather generally very bad this month; little frost, but dark, louring gloomy days, with excessive rain and sometimes terrible storms. The rain-fall extraordinary, and in consequence overflow of rivers and destructive floods in many parts of England, especially in the valley of the Thames; terrible mischief in the lower parts of London and at Windsor and elsewhere. Lakenheath fen flooded, and the Great Eastern Railway interrupted between Ely and Brandon. We were in much anxiety about Mildenhall Fen, but happily it has hitherto escaped. Here, being comparatively at a high level, 1875. we have not suffered at all.

———

A very pleasant party in the house this last week; our dear old friends the Arthur Herveys, with their one remaining unmarried daughter Caroline (Truey as they always call her)—Leopold and Lady Mary Powys, Lord John Hervey and his cousin John, The Dowager Lady Rayleigh, Mrs. Ellice and her daughter Helen; Captain and Lady Florence Barnardiston; our dear Minnie Napier; Edward and my nephew Harry (from the 3rd till this morning). Now we are quite alone, except Minnie.

The nick-name given to Mr. Ward Hunt, the First Lord of the Admiralty (in allusion to the frequent casualties lately in the navy)—Casual Ward. (Told me by L. Powys).

———

A fine, clear day, snow still lying deep.

Business with Scott. Lakenheath distress.

Read to Fanny and Minnie in the evening, some of Dean Hook's " Life of Beckett.

———

Harry returned from Cambridge.

We were much grieved by the news, some days ago, of the death of Mary Somerville. It happened at Castellamare.

1875.   I feel very much grieved for poor Martha, who is left, I believe, entirely alone in the world, without any near relations or connections. We have seen very little of the Somervilles for a long time past, but I used to know them well in old days, at Chelsea and afterwards at Rome, and Fanny knew them more intimately. When we last saw them, which was in '66, at La Spezia, Mrs. Somerville was still alive, though in extreme old age.

_____

December 14th.

Dear Minnie Napier went away. Very sorry we both were to part with her.

_____

December 15th.

Mr. Thornhill of Riddlesworth died on the 9th, aged only 71. Suppressed gout is said to have been the cause of his death. He was a fine specimen of the fine old English country gentleman.

_____

December 16th.

A tour of inspection with Fanny and Scott, through the groves and plantations.

We arranged some more books in the porch room.

_____

December 17th.

We had the delightful news that dear Kate Hoare has had a son, and that mother and child are doing well. I trust that her son will be a comfort to her and his father.

_____

Read to Fanny with much satisfaction, Lord Derby's excellent address to the students of the University of Edinburgh.

December 21st.

Began to read Green's "History of the English People." Little Charlie Seymour (Charles Hugh Napier Seymour), left with his nurse, to return to his father and mother, having been left so long under our care (ever since the 10th of November), on account of the danger of scarlet fever at Aldershot. He is an uncommonly well-grown and beautiful baby (just over a year old, and very amusing).

Another death in this strangely fatal year, which has made us very sorry,—that of Lady Sidney Osborne, an elder sister of Mrs. Kingsley. We saw her only twice—in London in June, 1871, and at Wells in April, '72, and neither time for more than a morning visit; but we were charmed with her, and earnestly wished for further opportunities of cultivating her acquaintance. She was very like Mrs. Kingsley in many respects.

December 22nd.

We went to Stowlangtoft and saw poor Mrs. Maitland Wilson.

December 23rd.

A very fine day.

1875.   Dear Katharine, Rosamond, Frank and Arthur
Lyell and Arthur MacMurdo arrived.

December 24th.

Showed Katharine our new greenhouses.

December 28th.

" *Hac data pœna diu viventibus—ut renovatâ.*
Semper clade domûs multis in luctibus, inque.
Perpetuo mœrore, et nigrâ veste senescant."

We are now near the end of this year. A sad
and fatal year, indeed, it has been to us, in a
remarkable degree : I do not remember to have ever
before, in any one year, lost so many, either real
and dear friends or pleasant acquaintances and
neighbours.   My dear brother Henry, Charles
Lyell, Harry Lyell, Charles Kingsley, Lady
Cullum, Lord Alfred and Lord Augustus Hervey,
Sir Edward Ryan, Maitland Wilson, Lady Sidney
Osborne, Emma Rowley, a melancholy list of losses.
Among them all there is hardly one whom I *miss*
more—whose departure I more often feel occasion
to regret—than Charles Kingsley.   In the case of
Henry, one can but feel thankful for his release
from hopeless suffering.   Charles Lyell's death, too,
we were quite prepared for : he had done his life's
work, and done it nobly.   But Kingsley was in the
full vigour of his intellectual faculties ; I never knew
him more delightful than he was the last time he
was here, in October '73, a short time before his

fatal expedition to America. Even after his return from thence, in September '74, when we saw him in London, though his health was evidently very much shaken, I perceived no symptom of decline in his mental powers.

Another in this list, the loss of whom we feel very often and of whom we are continually reminded, is our kind, hearty, genial, cheerful, constant friend and neighbour, Lady Cullum.

The want of Maitland Wilson is continually felt in the county business, and Sir Edward Ryan is both a public and a private loss.

Eminent men with whom I was not personally acquainted, who have died in the year:—Sir Arthur Helps, Lord Stanhope, Bishop Thirlwall, Dean Hook, Sir Francis Head, Sir Hope Grant, Professor Cairnes (a great political economist), Admiral Sherard Osborn, Comte de Jarnac, Lord St. Leonards.

Sir Arthur Helps I consider an especial loss.

But still, in spite of all these losses, I deeply feel that I have great reason to be thankful to the Almighty for the many blessings which I enjoy,— above all, that the greatest and most precious of them all is not threatened,—that my dear wife is fairly well in health, and (as I wrote last year), that we are as firmly knit in bonds of love and harmony as ever. I cannot feel too grateful for the many valuable friends who are still preserved to me, though many have been taken away. It has been a great delight to enjoy the society (though but for a short time) of Sarah Sanford and Kate Hoare and

1875. their father and mother,—to see Sarah's happiness
in her wedded life, and to find that her husband
is so worthy of her in every respect, and now, to
hear of the safety of Kate, and her happiness in
becoming a mother. The visit of the Albert
Seymours, too, and our short visit to them in
September, and their baby left so long in our
charge, these were very pleasant. Minnie is quite a
sister to us, and we cannot prize her too much.

I have noted the delightful glimpse which we
had of Rose Kingsley (September 29th), and I have
since had some very interesting letters both from
her and her mother.

All these incidents are very pleasant at the time,
and not less so to look back upon. As we go on
along the declining road of life, and friend after
friend is swept off, we feel the need of clinging
more closely than ever to those who remain. It is
cheering too, to see the children of our friends
growing up to fill places in our affection, and to
remind us of what their parents were in earlier days.

As to myself, the retrospect of the year is not
entirely satisfactory. I do not perceive any clear
indication of my having grown either wiser or better;
but old as I am, I hope that it is not yet too late for
improvement.

————

December 31st.

Miss Milner Gibson and her brother Gery came
to luncheon.

————

1876.

JOURNAL.

Had a very pleasant walk with Rose Kingsley. 1876. Mr. Bowyer went away.

Harry went back to Cambridge, and all our guests went away except dear Rose. I had a delightful walk with her.

LETTER.

Barton, Bury St. Edmund's,
February 3rd, '76.

My Dear Katharine,

We have had a very pleasant party, though our intended central figure (or protagonistes as the Greeks would have called him) failed us. Dr. Hooker had promised to come to us, with Miss Hooker, on Saturday the 29th January, to stay till the 1st of this month; Fanny took much pains to collect a party to meet them, and all came (except Dr. Marcet who had an attack of neuralgia), but no Dr. or Miss Hooker. It was rather ridiculous, as the different people success-

1876. fully arrived, who had been invited to luncheon or dinner to meet Dr. Hooker, to have to explain to them that there was no Dr. Hooker to meet. However, it was not his fault, for on Saturday he was deterred from travelling by the intense fog, and on Monday the Board of Works chose to make a swoop down on Kew for some official visitation, which he was obliged to attend. We could not blame him, but it was a great disappointment, for (besides other reasons), I particularly wished him to see our new houses. However, as I said, we had a very pleasant party, and above all, dear Rose Kingsley, who alone is amply sufficient to console us for any other deficiences and disappointments.

Our other guests were :—Mrs. Young and her daughter Caroline, and her son Charles (with whom you are acquainted); Mrs. Fuller Maitland, very good and kind, very clever, well informed and agreeable; Miss Maitland, very handsome, intelligent and pleasant; Mr. Loch (whom you know); Mr. Bowyer ; and two clever and well-informed clergymen of our neighbourhood, Mr. Grant and Mr. Lott, also my nephew Harry. Now all are gone except dear Rose, whom it is delightful to have staying with us.

To-morrow I shall be *sixty-seven*, a serious thought. I have great reason to be thankful for having been permitted to live to such an age in good health and with so many causes of happiness, especially so many good friends. Everybody says that Fanny is looking particularly well, and I think so too.

We have had very fine weather now for some 1876.
time, and those days when the fog seems to have
been so thick in London, were splendid here. I
hope it will be fine on Tuesday.

I am glad Leonard is with you. Pray give him
my love, as well as to Rosamond, and believe me

Ever your affectionate brother.

CHARLES J. F. BUNBURY.

JOURNAL.

Saturday 5th.

Yesterday was my 67th birthday. I have very
great cause for thankfulness, being in good health at
this age, and enjoying such a multitude of blessings ;
—above all such a wife, and such friends. For
these I can never be sufficiently grateful.

Monday, 7th.

Ground still covered with snow, which melted
slowly, though there was no frost.

Looked over some dried plants with Rose Kingsley. Rose Kingsley tells me, that bodies in the
keen winter weather of Canada, become so intensely
electrified, that she has herself seen a lady *light* a
jet of gas by merely applying her finger to it, after
shuffling with her feet on the floor, for a little while
to excite the electricity. And it is a common trick
of children in that country, to give an electric shock
to a stranger by merely touching him on the side
of the neck with a finger. She says that the

z

1876. oppression and discomfort produced by the rarified state of the air, in the excessively dry climate of the Rocky Mountains are much more severe than on the mountains of Europe at corresponding heights. She experienced great discomfort from this cause, at an elevation of only about 8000 feet. She says that her uncle, Dr. George Kingsley, who has seen almost every part of the world, declares the scenery of the " Yellowstone River " district (in North-west America) to be the finest and the most extraordinary that he has ever seen.

I read to the ladies in the evening, three of Coleridge's poems which are great favourites with me :—" The Hymn before Sunrise in the Valley of Chamounix," that entitled " Fears in Solitude,"— and the fine " Ode on France," (in 1797). All these to my thinking, are very noble—the first, magnificent. Rose, to whom they were new, fully shared my admiration.

LETTER.

Barton,
February, 10th, 1876.

My Dear Leonora,

I thank you very much for your kind and very pleasant letter of the 5th. I have, indeed, very great reason to be thankful for many blessings, and among them for my being in such good health at the age of 67, and able to derive so much enjoyment from study, from the observation of nature, and above all, from the company of real friends.

I am very glad to hear that Dora has a taste for 1876.
natural history. I hope she will keep it up, and she
will not want help and facilities for it in Germany.
I am glad to hear of Mr. and Mrs. Beyrich, and
also of Professor Ewald, who is, I suppose, the
geologist of that name who was very civil to me at
Berlin in 1855; he seemed then to be in bad health.
I have some specimens of fossil plants which he
gave me at that time.

Rose Kingsley has given me a valuable collection
of dried plants from California and Colorado,
collected by herself, the year before last; a great
many interesting things. I have also a few interesting plants from a high volcanic mountain in Java—
the same mountain which Wallace describes particularly — these gathered and given to me by
Dudley Hervey, one of Lord Charles's sons.

The next time you come here, you will see our
new greenhouses, which, as far as I can judge yet)
are very successful, and give me a great deal of
pleasure. They are in three divisions. The middle
(and highest) building is a hot-house ; one of the
wings for such plants as those of the Cape and
Australia, and South Europe ; the other for ferns.

Dr. Hooker has kindly sent us a number of
seedlings from Kew, for our hothouses, many of
them very curious and interesting. I hope they
will thrive with us. My love for botany is as great
as ever.

Prescott's " Peru " is indeed a very interesting
book, and beautifully written ; there is hardly
anything in modern history so exciting or of such

1876. romantic interest, as the narratives of these Spanish conquests in America, and nothing more shocking.

Did you ever read a work on the same subject by Sir Arthur Helps — "The Spanish Conquest of America." It is likewise extremely interesting, in some respects I think even better than Prescott, and, looking at the subject from a rather different point of view. From Helps's book we learn more of the efforts made on behalf of the poor natives, by good men among the Spaniards, especially by the monks, Las Casas and many others.

I have been reading to Fanny, in the evenings, Dean Hook's "Lives of the Archbishops of Canterbury," which is pleasant reading. We began it fully a year ago, but we have so seldom been alone, that our reading has been very intermittent.

I have lately read Virgil's "Georgics" through, to myself (not for the first time) ; exquisite poetry. I hope you have read, or will read the "Letters of Sarah Coleridge." I think them charming.

With my love to your family circle, believe me ever,

Your affectionate brother,
CHARLES J. F. BUNBURY.

February 12th.

JOURNAL.

Monday, 14th.

A heavy fall of snow in night, but day fine and not very cold. Dear Rose Kingsley left us: very

very sorry I was to part with her; at the same time I 1876.
quite appreciated her motives for leaving us. Her
sister's engagement, to be married to Mr. Harrison,
naturally makes her anxious to be at home again
(Mr. Harrison was formerly curate at Eversley, and
has long been a valued friend of the family, so that
the engagement is very satisfactory to them. Rose
received the announcement of it while staying with
us), and she has, first, to pay a long-promised visit
to the Fuller Maitlands at Stanstead. Rose has
been mostly alone with us since the party broke up
on the 3rd, and there could not be a more delightful inmate of our house. She is very like her father
in face, and not less so, I think, in intellect and
nobleness of mind, but with the additional charm of
feminine grace and sweetness.

Wednesday, 16th

A fine day.

Shocked and startled by the news this morning,
of the very sudden death of Mr. Fuller Maitland.
Rose Kingsley left us on Monday, to stay two or
three days with them (the Maitlands), at Stanstead :
we had a note from her yesterday morning,
announcing her safe arrival, and this morning
another letter from her, telling us that Mr. Maitland
was quite well on Monday evening (the 14th)—that
he got up at his usual hour, and apparently well
yesterday morning, and half-an-hour after, was
found lying quite dead in his dressing room.
Disease of the heart is supposed to have been the

1876. cause. I am very sorry. He was a very kind and good man, sensible, earnest and useful, well informed too, and not merely a practical man, but with a great love and knowledge of the fine arts, and a fine taste for them. He had especially studied the early Italian masters, and his house at Stanstead was rich in pictures of those schools, and likewise in English landscapes. It is curious that he should have resembled his friend, Sir Edmund Head, in the suddenness of his death.

We visited the Maitlands at Stanstead, in December, '72, and I then made a note in my journal of the pictures and books.

———

Friday, 18th.

Had a walk in garden with Fanny, and enjoyed the greenhouses.

———

Monday, 21st.

Fanny lost one of her beautiful little green paroquets (Melopsittacus undulatus), yesterday: it died quite suddenly. It is a great pity. She had four lovely little creatures, and rather tame.

———

Tuesday, 22nd.

A beautiful morning.

I see announced in the newspaper the death of Adolphe Brongniart, a great botanist, and the Linnæus of Vegetable Palæontology. He was very civil and pleasant to me when I was at Paris in 1847 and '57.

Fanny went yesterday morning to Livermere, to the wedding of Miss Horton, who was married to Mr. Oliphant, son of Colonel Oliphant, and nephew of Sir Antony Oliphant, who was Attorney General at the Cape when I was there, and first cousin of the more noted Laurence Oliphant, who was present at the wedding. I was not able to go to the wedding, by reason of a heavy disabling cold which has hung upon me for some time. Fanny says it was very well arranged, and passed off very pleasantly. They are going out immediately to India, where Mr. Oliphant has the appointment of Secretary to the famous Salar Jung, Prime Minister of the Deccan.

Thursday, 24th.

The East Suffolk election has terminated in the return of Col. Barne, the Conservative, by a majority of 950. I am well satisfied,—not that I know Colonel Barne, but he is a gentleman, and brother-in-law to Albert Seymour.

LETTER.

Barton,
February 25th, '76.

My Dear Joanna,

I thank you heartily for your kind wishes on my birthday, as well as for your very pleasant letter. I am, I assure you, very thankful to you and Susan for your good wishes and kind expressions, and it makes no sort of difference that they happened to arrive after the day intended; the

1876. value of my friends' good-will is not so evanescent.
I am glad that you have been spending a pleasant
winter.

I am reading Miss Backley's "History of Natural
Sciences," and think it very well done : it is very
much condemned of course, but (as far as I have
gone), is clear and well arranged, and the facts
judiciously selected. Though I have not yet read
further than the middle of the 16th century, I have
already learnt from it some things which I did not
know. Her dedication to the memory of Charles
and Mary Lyell is excellent, good and warm feeling
gracefully expressed. I am reading also Green's
"Short History of the English People:" another
multum in parvo book, very unpleasing in outward
appearance, and printed in horridly close type,
somewhat fatiguing to read, too, on account of its
extreme conciseness and condensation, but well
worth reading. It is full of matter very vigorously
written, and some of the more striking incidents told
with great spirit. It was recommended to me by
Mr. Sanford (Sarah Hervey's husband), who is a
remarkably accomplished and well-read man.   I
have learnt much from it, but, indeed, my
knowledge of the mediæval history of England was
very scanty indeed. We have both been reading a
very odd sort of novel called "For Sceptre and
Crown," translated from the German ; but I am at
the end of my paper and must leave off.

With much love to Susan,

Ever your affectionate brother,

CHARLES J. F. BUNBURY.

JOURNAL.

Up to London.

A short visit in evening from Katharine and Rosamond.

The Spanish news remarkable; the Carlist insurrection has collapsed at last rather suddenly; Don Carlos has retired into France, and is said to be coming to England: his followers are laying down their arms in mass: the war appears to be over, unless it degenerates into mere brigandage.— But the Government of Spain, after the cessation of this war, will be anything but an easy problem.

Much rain.   48, Eaton Place.

Clement breakfasted with us.   Minnie drove with us: visited Sarah Sanford, but she was not well, and Mr. Sanford out,—then to Lady Lilford's, and we had a pleasant talk with Minnie Powys.

Minnie (Napier) dined with us.

We and Minnie went to Burlington House and saw the pictures.   The gem of the whole exhibition, to my mind, is *Romney's* "Lady Hamilton at a Spinning Wheel," one of the most fascinating pictures—almost the lovliest face I ever saw.

1876. Looking at this, one can well believe all that is told
of the fascinations of Lady Hamilton. Other
pictures which delighted me much were:—" The
Cottage Girl," by *Gainsborough;* " Portrait of
Garrick," by *Gainsborough;* Group of Portraits of
Five Children of Marqius of Stafford, by *Romney;*
" Cupids Harvesting, by *Rubens;* " Garrick and his
Wife," by *Hogarth;* " Mrs. Bouverie and her
Child," by *Sir Joshua Reynolds*—(charming).

Lord Normanton is the fortunate owner of
Romney's " Lady Hamilton."

Friday, 3rd.

We visited the Miss Moores. Mr. and Miss
Gibson Craig visited us. Again with Minnie to the
Old Masters at Burlington House; enjoyed it much.
Thought Lady Hamilton as irresistibly fascinating
as ever. Noticed also with pleasure, some ad-
ditional pictures:—*Reynolds's* "Mrs. Abington as
Miss Prue" (lent by Lord Morley—capital). *Hogarth's*
"Peg Woffington." *Gainsborough's* "Pomeranian
Dog and Puppy"—(delightful); even superior, I
think, to most of Landseer's dogs); portrait of Hon.
Laurence Bouverie, with two dogs, by *Cosway;* and
portrait of Hon. Frances Harris (Lady Frances
Cole) as a child with a dog, by *Sir Joshua.* The
three portraits of Garrick, by *Sir Joshua, Gainsborough, and Dance,* placed near together, are
interesting to compare. A large landscape with the
Escurial in the middle distance is a very curious,
peculiar, and interesting picture. It is said to be

by "one *Verhulst*, an artist of Antwerp," and to 1876.
have been "finished" by *Rubens*. It appears very
unlike *Rubens's* usual manner and style of colouring;
there is something strangely *weird* and spectral
in the effect of light and colour, as if the landscape
were seen during an eclipse; and this seems to
harmonize very well with the wild character of the
mountain scenery, as well as with my pre-conceived
idea of the Escurial.

Saturday, 4th.

We drove in Battersea Park, and back by the
embankment past Chelsea.

Minnie and her brother, John Herbert, Edward
and Clement dined with us.

Monday, 6th.

Mr. S. Smith, Mrs. Coltman, and dear Rose
Kingsley, came to luncheon with us ; afterwards we
took Rose to the Sanfords—saw Sarah, Mr. Sanford
and Nellie. Sarah is confined to her sofa, and
looking *terribly* thin and pale, but as cordial and as
animated as ever. Mr. Sanford came in from the
drawing room in his uniform, and greeted us with
great cordiality.

Tuesday 7th.

Wrote to Cissy.

A visit from Mr. Sanford—very agreeable. I like
him more and more. He talked of mountain

1876. climbing, and gave me very interesting accounts
of his adventures in the Alps, hazardous climbs
on the glaciers of the Oberland, passing over
narrow bridges of ice, with showers of stones falling
and threatening to sweep him and his guides away,
and such like hair-breadth escapes.

We dined with Minnie; the company, her brother
John, Sarah and Albert Seymour, Lady Horatia
Erskine, Miss Macdonald, and Richard Strutt; a
very pleasant little party. Sarah looking very well,
and as pleasant as ever. Lady Horatia very
handsome, and very pleasing.

_____

Thursday, 9th.

In the morning a *very* pleasant visit from dear
Kate Hoare ; also one from Margaret Richmond
(Aberdare's eldest daughter), with her beautiful
little boy. In the afternoon, Mr. Walrond, also
very pleasant ; he had come up from Bournemouth
on purpose to attend the funeral of Lady Augusta
Stanley :—a very fine and solemn scene, he said,
especially when occasional gleams of sunlight fell on
the "sea of heads." At luncheon, Norah Aberdare
and William Napier, two people whom I both like
and esteem in a very high degree.

Kate tells me that Canon Beadon of Wells is
reported to be in an excellent state both of body and
mind. He will be 99 next December, it is said.

William Napier entirely approves of Mr. Ga-
thorne Hardy's proposed measures for the stength-
ening of the army, as explained in his speech
on the Estimates.

Charles Mallet breakfasted with us ; we were very glad to hear from him that he had just received a letter from his brother Louis from Bologna, with a good account of himself ; it is a satisfaction to know that Louis is safe in Europe again, and clear of the hot and unhealthy climates. The disease appears to have quite left him, but he is still very weak. His illness appears to have been very serious indeed.

Saturday, 11th,

Wo went to Lady Mary Egerton's, at Mountfield in Sussex, by the 1.25 train on the Hastings line— Mrs. Clements Markham with us. Mr. Clements Markham and Sir Francis Doyle at Mountfield— both agreeable.

Monday, 13th.

We spent the morning very pleassntly at Mountfield—returned to London by the 2.40 train from Robertsbridge. Sir Francis Doyle with us ; very agreeable. We arrived safe in Eaton Place. Thank God.

Sir Francis Doyle is exceedingly pleasant, with quiet, gentle manners and excellent conversation ; literature and especially poetry is of course his *specialité* (he is Professor of Poetry at Oxford) but he has extensive knowledge both of books and men, and abundance of agreeable anecdote.

He spoke favourably of Swinburne's last work, "Erechtheus." Sir Francis thinks that Keat's

1876. death was the greatest loss that English poetry has sustained for a long time — that considering how young he died—Keats' poetry was really extraordinary.

The good news that the Prince of Wales has actually started from Bombay on his homeward voyage—so he is safe out of India.

Tuesday, 14th.

Fine, but very cold and windy. We had luncheon with the Aberdares.

Sir Frederick and Lady Grey came to stay with us—Sir Frederick dined out,—Lady Grey and the Charles Hoares dined with us—very pleasant.

Wednesday, 15th.

Our dinner party: the Bishop of Bath and Wells, Sir Frederick and Lady Grey, Rose Kingsley, Mr. Sanford and his daughter Nellie, Minnie Powys, Minnie Napier, Katharine, William Napier, Mr. White (preacher at the Savoy), General and Mrs. Rumley, and Edward—fourteen besides ourselves, and all pleasant. A few more in the evening.

Thursday, 16th.

Sir Frederick and Lady Grey, came to us on the 14th, from Lynwood, to dine with us on the 15th ; dear Rose Kingsley came to us yesterday, only in time to dress for dinner.

We had a very pleasant breakfast with them ; but

they left us this afternoon. The Greys very agreeable ; Rose charming as she always is.

Much talk about the " Royal Titles." We are all
against the assumption of this new-fangled title of
Empress ; Sir Frederick vehement against it, and
he says he has hardly met with any one among all
those with whom he has talked, who approves of it.
He said, and I heartily agree with him, that there
cannot be a higher title. nor one of more real
dignity than that of Queen of England. Unfortunately, he says, it is believed that this new title is a
personal fancy of the Queen herself. I thought she
was wiser.

Friday, 17th.

The Ministry, I see, carried their Royal Titles
Bill by a large majority, 305 to 200; but it was no
doubt a regular party vote.

March 18th,

Looking over the lists of the division, I do not
recognize any Whig names in the majority, but
probably many Whigs abstained from voting.

We dined yesterday with the Sanfords, met the
Bishop of Bath and Wells, and the Charles Hoares,
Sir Henry and Lady Rawlinson, Professor and Mrs.
Flower, Mrs. Harvey (of Ickwellbury) and Miss
Harvey, Mr. Seymour (a brother-in-law of Mr.
Sanford) William Hoare. Dear Sarah, still a
prisoner, and not able to go down to dinner with us

1876. but looking much better than when I saw her the other day, and very cheerful. I sat by Kate, who is looking brilliantly well, and had exceedingly pleasant talk with her.

— — —

Sunday, 19th.

A heavy fall of snow in the morning, the snow lying in the streets till late.

Our second dinner party was yesterday; Norah Aberdare, the Charles Hoares, the Clements Markhams, Mrs. Edward Romilly, Dr. Marcet, Mr. Bentham, Caroline *(Truey)* Hervey, Rosamond Lyell, Mr. and Mrs. Fazakerly, William Hoare and Edward—all very pleasant. In the evening, Mr. Sanford and Nellie, Minnie Napier, Katharine, Admiral Spencer and a few others. A very pretty bevy of young ladies : Rosamond, Caroline Hervey, Nellie Sanford and Miss Newton.

At dinner I sat by Norah Aberdare, who is, under such circumstances remarkably agreeable, besides being always admirable both for greatness of mind and for goodness.

Norah's account of the distressed state of the people in her district, about Aberdare (in Glamorganshire), owing to the effects of the great *strike* a year or two ago ; the injury done by the strike still very generally felt ; the iron trade, in particular, almost ruined, so that she doubts whether it will ever recover. The prosperity of the district much affected also by Lord Bate's difficulties.

— — —

We called on and saw Sir Edward and Lady Blackett—afterwards went to luncheon with Lady Lilford. Miss Elizabeth Powys there as well as Minnie—also a Mr. Sinclair. Fanny went to the Sanford's evening party.

Mrs. Ellice, Mrs. H. Grenfell, the Bishop of Bath and Wells and Minnie came to luncheon—very pleasant.

The MacMurdos dined with us, just returned from Italy.

William Napier breakfasted with us.

Rose Kingsley, Lady Charlotte Legge, Minnie and Mr. Goodlake came to luncheon.

Fanny went out driving with Rose and Minnie. A visit from Edward.

Weather rather improved.

I walked out, first time this month, called on Carrick Moore and had a good talk with him.

Minnie and her brother dined with us.

Sarah Seymour, her mother, Katharine and Rosamond and Kate Hoare came to luncheon—all very pleasant.

1876.   We dined with the Charles Hoares, met the Bishop and the Paleys.

Saturday, 25th.

We returned home, by the noon-day train from the new Liverpool Street Station : a good and convenient train which brought us as far as Marks Tey in little more than an hour. We returned home ourselves in good health, and found all safe and well at our dear home—thank God.

Tuesday, 28th.

A delightful change of weather—quite spring. Had a pleasant little drive in the pony carriage with Fanny, and a stroll afterwards. Mr. Percy Smith came to luncheon.

Wednesday, 29th.

A walk in the garden with Fanny.

Friday, 31st.

A beautiful spring day.

We walked through the Vicarage Grove, directed cutting of laurels, afterwards drove round the park in pony carriage.

I have not yet noted the recent re-marriages of two of our friends :—Mr. Walrond to Miss Louisa Grenfell, and Mr. Hutchings to Miss Farquharson of Invercauld.

Edward, dining with the Edward Hamiltons on 1876. the 21st, met Gladstone and John Bright. He told me he was much struck with the great change in Bright's manner and demeanour since the days when he (Edward) used to meet him in the House of Commons:—that he has grown so much more gentle, quiet and unobtrusive in manner, not talking much, rather silent, and speaking in a low voice and very quietly when he does talk. Gladstone, he said, talked much, and was extremely agreeable.

_____

Saturday, April 1st.

A very fine day, but with a cold wind. We were out of doors most of the morning, looking round the groves and shrubberies with Scott, and walking about the arboretum.

_____

Sunday, 2nd.

We went to morning Church, and received the Communion. First time I had been at Church since November.

_____

Tuesday, 4th.

We drove round by Lofts's farm and Conyer's Green. Also visited the new plant houses.

_____

Sunday, 9th.

Yesterday we went to Hardwick. I had not been there since Lady Cullum's death. It was a great

1876. though a melancholy pleasure to go over all the spots
which we had so often visited with our dear, kind,
genial friend, and which she took so much delight in
showing. Everything recalled her to mind: one
seemed to see and hear her at every turn. Poor
Mr. Fish, who was quite devoted to her, and whom
she always treated as a real friend, feels very
severely the changed state of things and the lone-
liness of Hardwick. He seemed to take pleasure in
accompanying us through the plant houses and the
arboretum, &c., talking with us about the plants,
and speaking of Lady Cullum. The gardens and
all about the place are in admirable order. I have
long thought it the most beautiful place in all this
part of Suffolk, and it seems likely to keep this
character.

We hardly looked into the house, which is un-
occupied except by servants.

The day was brilliant, and Hardwick appeared in
all its beauty ; all alive with birds and bright with
spring flowers and fresh bursting leaves.

———

Tuesday, 11th.

The 32nd anniversary of our happy engagement.
Mrs. Rennie returned from London, and Louisa
MacMurdo with her.

Clement came later.

———

Wednesday, 12th.

Arthur arrived from School. Wrote some geological notes on Mr. Horner's MS. for Fanny.

The weather is extraordinary. From March 28th to the 9th of this month, it was beautiful and delightful : bright, warm, and genial, all that one could wish Spring weather to be ; some days perfect summer. The 8th was especially fine. The 9th rough and stormy, the 10th worse, with violent storms of rain, of which we had hitherto had none this month.

The 11th plunged us back again into the middle of winter ; a succession of furious storms of sleet and actual snow, which whitened the ground, though it did not lie long. The cold is excessive, and this morning there was actually a heavy fall of snow, but it melted before noon.

———

Friday, 14th.

Read Kingsley's fine sermon on The Perfect Love, and read St. John, chap. xix. in Greek.

———

Friday, 21st.

Clement went away.

The news of Lord Lyttleton's melancholy death.

I did not know him, but should think that he was a man to be much regretted.

We heard a few days ago of Louis Mallet's safe arrival in London ; very welcome news.

———

Saturday, 22nd.

Kate Hoare and her husband and Nellie Sanford arrived, also Harry.

We heard the cuckoo.

1876.   Sunday, 23rd.

We all went to morning Church, and afterwards walked about the gardens.

_____

Monday, 24th.

Received notice of my election into Royal Botanic Society.

_____

Tuesday, 25th.

Katharine and Rosamond arrived; also Captain and Lady Florence Barnardiston.

Fanny and almost all the party went to Mrs. Brown's ball.

_____

Wednesday, 26th.

Showed some of my minerals to Charley Napier; afterwards a very pleasant stroll through the gardens with Katharine, Rosamond, and Lady Florence.

Our dance.

_____

Thursday, 27th.

William and Charlie Napier, Charles Hoare, the Barnardistons, and John Hervey went away. Lord and Lady Hanmer, Mr. and Lady Mary Phipps, and Mr. Lott arrived.

_____

Saturday, 29th.

Fanny took Lord Hanmer over to Mildenhall, returning at six.

_____

Monday, May 1st.

Lord and Lady Hanmer went away—also Harry, Arthur and his sister alone remaining with us.

Tuesday, 2nd.

Very pleasant company in our house last week. Lord and Lady Hanmer were here for the first time, though we have known them for some time in London. They are both very cheerful, good-natured and pleasant, having especially that great ingredient of pleasantness, the being easily pleased. *He* has a good deal of taste and knowledge.

Wednesday, 3rd.

Went to the Jail Committee at Bury :—met there Barnardiston, Joseph Rowley and Mr. Young—very few prisoners in the Jail.

Colonel Marshall (Suffolk Militia) came to luncheon.

Thursday, 4th.

Dear Arthur went back to school.

Saturday, 6th.

Fanny went to Mildenhall to see Mrs. Marr.

Monday, 8th.

Mr. James of Livermere came to luncheon : a serious talk with him afterwards—a wise man.

Wednesday, 10th.

Up to London.

Visits from the MacMurdos and Edward. Minnie dined with us.

1876.   Thursday, 11th.

The Prince of Wales' entry into London, on his
return from India.   It is a great satisfaction to
know that he is really returned home safe and
sound from so long an expedition, after so many
anxieties and apprehensions and so many prophecies
of evil.

Dear Rose Kingsley came to stay with us.   She
and Fanny went to a balcony at Mrs. Horton's, in
Grosvenor Place, to see the Prince pass.   They had
a waiting job, for he did not come till near 7, but
they saw him well.

———————

Friday, 12th.

Visit from Isabel Hervey.   Fanny took me in the
carriage to the Athenæum—met Lord Aberdare
there.

Minnie and Edward dined with us.

———————

Saturday, 13th.

Called on Lady Augusta Seymour and on John
Moore—both pleasant.

———————

Sunday, 14th.

Read prayers with Fanny and Rose.

Visits from the Hanmers, Mr. and Mrs.
Hutchings, Mr. Bentham, Fred and Jack Freeman
and Clement.

———————

Monday, 15th.

Susan MacMurdo, Minnie Powys, Lady Charlotte
and Lady Octavia Legge, and *our* Minnie came to
luncheon.

The Albert Seymours, May Egerton, Minnie and 1876.
Fred Freeman dined with us.

Susan MacMurdo is one of the handsomest
women, of her age, I have ever seen (I believe she
is turned of 50). Minnie Powys very pretty, looking
remarkably well. Sarah Seymour also looking very
well and very engaging.

————

Tuesday, 16th.

Rose Kingsley's talk about what she has seen in
the distant countries she has visited—West Indies,
North America and Mexico—is delightful. She is
like her father (as I think I have mentioned before),
in many things, and, among others, eminently so in
acuteness of observation and in power of describing
what she has seen. I can in no way do her
justice by any notes of what I have heard from
her. She has, indeed, a good deal of her father's
felicity of expression.

[Mem.—Her observations on thunderstorms,—
frequency of them among the Rocky Mountains and
on the plains near: great thunderstorms on summits
of Pike's Peak, raging for nearly 12 hours,—misery
of the surveying party encamped there, who were
in the midst of it the whole time, and were near
going mad from excited sensations: the telegraph
wires melted.—Railway trains in motion rarely
struck by lightning—why?—one instance in
America.

The autumn much the most beautiful season of
the year in America.

————

1876.   Wednesday, 17th.

A visit from Cecil.

At the Athenæum, saw Blakesley and Alfred Newton.

Rose dined out.   Clement dined with us.

———

Thursday, 18th.

An early visit from Douglas Galton, who is going to Philadelphia to the Centennial Exhibition. Talked about the Exhibition of Scientific Instruments lately opened; he thought the arrangement imperfect, and that there ought to be explanatory labels instead of mere references to a catalogue.

Lord Napier* came to luncheon.   I had not seen him since he was an undergraduate at Cambridge. We found him very agreeable, lively, cheerful and accomplished: much interested in pictures and in family histories.   I did not perceive in him any striking likeness, either in face or manners (except that he has a large nose), to those Napiers whom I have known so well (the children and grand-children of Col. George and Lady Sarah; but he is a good looking man.

After luncheon we went (Fanny, Rose, Minnie Napier, Minnie Powys and I), to Rose Bank, and spent the afternoon pleasantly with the MacMurdos. They have made their house and their conservatory very pretty.   Susan has a remarkable share of taste in all the decorative arts.   Their garden was overflowed by the river in the course of the winter, and is only now beginning to recover.

* Ettrick.

Mimi MacMurdo's drawings, especially of flowers, 1876. are really beautiful. She is, herself, an un-commonly pretty girl, very clever, and every way attractive.

———

Katharine and Rosamond came to luncheon.

We went to a crowded afternoon party at Mrs. H. Berners' to hear "Corney Grain."

———

Fanny and Rose, with Minnie Napier and Minnie Powys, went to the Freres at Wimbledon.

Called on Sally, and walked in St. James' Park.

We went to the Geographical soireé at the India Museum. There was a great number of people, but the great gallery of the Museum afforded so much space, and in so convenient a form, that there was no crowd. I had not seen that Museum before. The great gallery, as I have said, appears very handsome and conveniently proportioned, and the profusion of Oriental objects ranged along both sides of it would, I daresay, be worth looking at more carefully. In a narrower corridor, through which we passed first, were hung a great many large maps, which also would no doubt have been worth studying. Sir H. Rawlinson pointed out one as Lieut. Cameron's original map, on which he had marked down all his discoveries, but I could not see it well. We met many whom we knew.—The

1876. MacMurdos, Lady Mary Egerton and two of her
daughters, George Bentham, John Evans (late
President of the G. S.), Lord and Lady Hanmer.

———

Sunday, 21st.

Fanny went with Rose in morning to the Savoy:
in afternoon with me to St. Peters: later we called
on and saw Mrs. Grey and Miss Shirreff and the
Egertons.

———

Tuesday, 23rd.

We (Fanny and I, without Rose), dined with the
Hugh Bernerses: met the Charles Hoares, Admiral
Spencer, Admiral and Mrs. Sotheby, Mr. and Mrs.
Rycroft and a few others—a pleasant party. I sat
by Kate Hoare—charming. Mrs. Rycroft (Maud
Berners) is very pleasant.

A visit from dear Sarah Sanford—delightful as
ever.

———

Wednesday, 24th.

Our dinner party:—Louis and Fanny Mallet,
Georgie Pellew, Lady Mary Egerton, Minnie
Powys, Lady Head, Mr. and Mrs. Godfrey
Lushington, Mr. Marcet, Mr. Bentham, Montagu
MacMurdo, Mr. Rutson, Clement, besides Rose and
ourselves. A very pleasant party.

Louis Mallet seems quite recovered and in good
spirits.

———

Edward called, and brought us the first news of the strange theft of the portrait of the Duchess of Devonshire, supposed to be by Gainsborough, and lately bought by a picture-dealer for £10,000,—how it had been cut out of the frame and carried off from the room where it was exhibited.

Our 2nd dinner party:—Lady Louisa Legge, Lady Barbara and Mr. Yeatman, Sarah and Albert Seymour, Minnie, Kate and Charles Hoare, Mr. and Mrs. Rycroft, Dowager Lady Rayleigh, Admiral Spencer, Mr. Harrison, John Herbert—a pleasant party.

————

A pleasant visit from Lord Napier.

We dined with the Hanmers:—met Lord and Lady George Cavendish, Lord Crewe, Lord and Lady Tollemache, Mrs. Dugdale, Admiral and Mrs. Eden and others.

I was glad to renew acquaintance with my old college friend, Lord George Cavendish, whom I had known very well at Trinity when we were both undergraduates, but (as it happened) had never met since. He was very cordial and pleasant. In those Trinity days he was an excellent fellow in every sense, and I believe he is a very good man. He has a very good countenance. He told me that he had now been in the House of Commons for above 40 years.

I sat by Mrs. Dugdale, whose son has married

1876. one of the Miss Trevelyans, a niece of Macaulay. We had a great deal of talk about the " Life," and sympathized heartily in admiration of Macaulay and delight in the book.   She said it was quite true that, as far as is known, he never was in love ; that his nieces say they do not remember ever to have heard any story or tradition of a love affair of his,—not even a joke on the subject.

———— —

Monday, 29th.

A very fine day.

We went to Kew, to a garden party, invited by Joseph Hooker and his daughter (a pretty girl), at his house, where there is a little garden opening into the great gardens.   There were a great many people—many "distinguished foreigners" it was said, but we knew none of them, and there was nobody to tell us who they were.   We met the Macs, and walked with them about the great gardens.   These are now in great beauty and very delightful : the great trees in their lovely drapery of fresh young leaves, the grass brilliantly green, the flowering shrubs in their beauty, the Rhododendrons particularly, in great masses of blossom of many various shades of white, pink, purple and crimson, very beautiful.

————

Tuesday, 30th.

A beautiful day.   The 32nd anniversary of our happy wedding ; thanks be to God for all the blessings He has bestowed on me, and for this especially.

My first acquaintance with Kate Hoare's beautiful 1876.
baby.

The news of the deposition of the Sultan. A
revolution very quietly effected, hitherto so far as
appears. I wish it may tend to a settlement of the
formidable "Eastern question," and the preservation
of peace.

We had a snug little dinner party of seven.
Sarah and Mr. Sanford and Nellie, Minnie Napier,
Rose, and our two selves. Exceedingly pleasant.

Wednesday, 31st.

Dear little Charlie Seymour was brought to see
us.

We went to an afternoon party at Mrs. Horton's
in Grosvenor Place—met many we knew.

June 1st (Thursday).

I went to Linnean Society meeting at Burlington
House.

Friday, 2nd.

Louis Mallet came to luncheon. I had a long
talk with him about India. His opinion about the
disputed questions between the late Viceroy and
Lord Salisbury. Lord Northbrook is an old friend
of Mallet, yet he (M.) thinks him quite mistaken in
his *tariff* policy. Lord Salisbury's first dispatch
calm and temperate; unluckily, Lord Northbrook
had passed his tariff act before he received it.
That act a surprise; passed at Simla (where legis-

1876. lative business is seldom transacted), at an unusual season, and with an appearance of haste. Lord Salisbury thought it was an intentional defiance, and was angry, and wrote accordingly. Mallet was sent out to explain matters, and try to reconcile the two, but had hardly arrived at Calcutta before he was disabled by illness. The worst part of the matter, the conflict of authorities, which has been introduced into it, especially by the press and the Parliament. Mallet seems to hold that the Government of India ought to be carried on in subordination to, and under the immediate control of, the the British Government; and, as he himself acknowledged, the "British Government" now means the House of Commons. I objected the danger of governing India according to the notions of a majority ignorant of Indian affairs, and perhaps according to English rather than to Indian interests. He did not think much of this, because the Council of India (in London), without which the Secretary for India does nothing, is composed of men of old Indian experience, quite as well acquainted with the interests and habits and wants of the people of India as the Council at Calcutta ever are.

I have a very great respect for Louis Mallet, and for his opinions; but on this matter I cannot help having doubts.

———

Saturday, 3rd.

We, Fanny and Rose and I, with the Sanfords (Sarah and her husband, and his four daughters), went to the Zoological Gardens, and spent a very interesting and pleasant afternooon there, though 1876. there was a great crowd. Mr. Sanford has such an extensive knowledge of animals, such a love of observing them, and such animation in talking of natural history, that he is a particularly interesting and agreeable companion in such a visit. I like him more and more as I know him better. His daughters are very nice girls. He is also well acquainted with the keepers, and thus can get objects shown which are not seen by the common run of visitors. Thus, one of them roused from his slumbers, the Cape Ant-eater, Orycteropus, a great rarity :—let us see him for a little while ; an ugly beast he is, at first sight not unlike an ill-shaped hog, with large ears and a long snout, but by no means so extraordinary as the real ant-eater. Of *that* strangest looking of all beasts we had a good view ; he did not seem to be nearly so recluse as his Cape kinsman ; and the keeper showed us, to my great surprise, that this ant-eater is both able and willing to swallow mice. Other things we saw were the young Elephants brought from India by the Prince—delightful little creatures, really not higher than a pony, most tame, docile, and sagacious. The "Clouded Tiger" from Burmah, a very beautiful species of leopard ; this in the new "Lion House," which is certainly a great improvement on the old narrow dens for the carnivora ; and here there are two or three beautiful Jaguars, as well as several fine Tigers, Lions, and Leopards. A "Mouflon," female, I suppose, and a young one ; more like a goat than a sheep. A wild Ram from

1876. the mountains of the Punjaub, with magnificent
horns. The *Binturong*, a very singular creature,
with thick fur of a dark grey colour, and a long,
bushy, curly tail ; Mr. Sanford says it is a fru-
givorous " Bear," living in trees, and having a
prehensile tail.

Moose Deer, from North America ; their horns
yet in a young stage, but otherwise they look very
like the pictures of the North European Elk, and
Mr. Sanford said he believed them to be the same.

Monday, 5th.

Katharine, Rosamond, Minnie, Sarah, Sally, John
Herbert, and Edward, dined with us.

LETTER.

48, Eaton Place, S.W.

June 5th, 76.

My dearest Cissy,

Fanny tells me you wish to know what
I think of political affairs. I can only answer as a
philosopher I have heard of replied to a puzzling question :— "*Ma foi, Madame, Je ne sais
qu'en penser.*"

I have observed that political prognostics, made
with the greatest confidence, and even by men of
great note, are so very apt to be falsified by the
event, that I have determined not to attempt anything of the sort ; and I do not remember a time
when the political look-out appeared more confused

or uncertain, or anything like prophecy more 1876.
hazardous.

The events of the last few days at Constantinople
have been startling and sensational enough in all
conscience.

Do you believe that the suicide of the ex-Sultan
was quite unassisted? Deposed Oriental sovereigns
have been very apt to find benevolent friends to
help them out of this troublesome world. And what
next? Can we trust to the new Sultan's promises
of reform any more than to former ones? For
my part, I sympathize with the insurgents, and wish
that all the Christian populations subject to Turkey
may become actually (though perhaps not nomi-
nally), independent. But the question is terribly
complicated by the interests and ambitions of
rulers as well as by the passions of the peoples. I
earnestly hope that we shall not be drawn into a
war; a war to support Turkey in oppressing its
dependencies would be atrocious. I fancy that
most people now think that our former Russian war
was a great mistake, and that we were made the
catspaw of Emperor Nap. I have a good deal of
confidence in Lord Derby; he is not likely to be
rash or hasty, and I have hopes that he will not let
us *drift* into war as Lord Aberdeen did.

June 8th.—I have kept back my letter in hopes of
hearing something that might be worth telling you.
But I hear nothing of any consequence. In fact,
you may see by the newspapers themselves how
uncertain and unsettled all people are, except

1876. perhaps those few who are in the secret—if there be a secret.

I still trust there will be no war. Yesterday at dinner at the Nugents we met Lord Cathcart, whose father was one of my father's old friends and fellow-soldiers in Sicily. He (Lord Cathcart) does not think it probable there will be war ; because, he says, it is not the interest or the wish of any one of the great powers, except Russia, and she will be held back by want of money, as well as by the character of the Emperor. Like me too he has confidence in Lord Derby, and thinks he will not repeat the mistake of Lord Aberdeen.

Addio. Fanny has been at large evening parties, two nights running, and is much fagged. We have both got coughs, which though slight, we cannot yet shake off. Give my best love to dear Emmie.

Ever your truly loving brother,

CHARLES J. F. BUNBURY.

JOURNAL.

Tuesday, 6th.

We (three) dined with the Charles Hoares—met Lady Acland, Dow. Lady Rayleigh, Mr. and Mrs. Bouverie, Caroline Hervey, Captain Medlycott, and others; after dinner came Sarah and Nellie Sanford and a few more. Fanny and Rose went to Lady Burdett Coutts' party.

The startling news of the suicide of the ex-Sultan Abdul-Aziz. I never read or heard before of any Mahommedan killing himself. I thought it was quite inconsistent with all their habits as well as 1876. their beliefs. I wonder whether this suicide was quite unassisted.

Wednesday, 7th.

Mrs. Martineau and Mr. Medlycott came to luncheon.

Visit from Mr. and Mrs. Drummond. We (two) dined with Sir George and Lady Nugent; met Lord and Lady Cathcart, Sir Francis and Lady Boileau, and others.

Fanny and Rose went to Mrs. Goodlake's party.

Thursday, 8th.

We three went, with the Sanfords, Katharine and Rosamond, and Harry Bruce, to Rose Bank, and spent the afternoon with the MacMurdos very pleasantly—many people there—many whom we knew. The day was fine, even warm (till towards evening) ; the river full (luckily "without o'er-flowing"), and Rose Bank looked very pretty. The garden has been restored from the effects of the flood. I need hardly say that the MacMurdos were very cordial and very pleasant. There was a large party, many of our friends and acquaintances : — (Kate Hoare and her sister Caroline—Truey) ; Lady Head and her daughter Amabel ; Lady Mary Egerton and May and Charlotte ; Sir William Codrington ; Captain Horton and Freda Broke. One of the guests, Mr. MacCullum, an artist, had brought with him a beautiful dog from the East,

a Persian greyhound, very remarkable; like a
common greyhound in general form, but the ears
and tail fringed with long hair, as in a spaniel;
otherwise smooth-haired; the colour fawn. It was
very gentle and very affectionate to its master. We
saw this dog again in a visit to Mr. MacCallum's
house at Kensington. It is remarkably gentle and
good-natured.

I was very glad to see dear Sarah Sanford quite
restored (apparently) to health, and to her former
gay and lively spirits.

————

Saturday, 10th.

We went (Fanny and I), to the Royal Academy;
had the pleasure of meeting Sarah and Mr. Sanford
there—Constantine Hervey with them. The picture
which pleased me by far the most of all in the
Exhibition, and which is quite delightful, is *Millais's*
landscape, "Over the hills and far away." I prefer
it to anything else whatever of *Millais's* that I have
seen. It is a moorland scene, may be on the skirts
of the highlands; a foreground of heath and bog;
low rocky hills on either hand in the middle distance; and in the background a gradually ascending
valley, losing itself among hills becoming blue in
the distance. The truth of the details, combined
with the admirable general effect of the whole, are
most remarkable. Rose tells me that she knows
the place; that it is a moor about six miles from
Dunkeld, in the direction of Blair Athol. (My
description is not quite correct; beyond the rocky

hills in the middle distance is a somewhat open 1876. valley with a river (the Tay ? or the Garry ?) seemingly at a lower level than the foreground ; and the blue hills in the distance are the mountains about Blair Athol.

There is a very beautiful landscape by *Vicat Cole* —" The Day's Decline." A rich glowing English scene, in the style characteristic of him. A remarkable picture of the Carrara marble quarries by *Poingdestre*—very clever, and (as it appears to me), very true.

"The Bombardment of the Acropolis at Athens by the Turks," by *A. MacCallum ;* * a striking and impressive picture ; remarkable effect of light.† A lovely portrait of Mrs. Schlesinger, by *Millais.* Captain Burton (the traveller) by *Leighton*—very vigorous. A portrait of Mr. Bouverie (the Rt. Hon.) by *Ouless ;* powerfully painted, but in a style a little inclining to caricature.

Goodall's "Holy Mother" is in some respects beautiful, but its treatment of such a subject is not, to my thinking, quite in good taste. The costume is, I dare say, correct as showing what is now worn by the maidens of that country, but it appears to me indecorous for the Virgin Mary.

Monday, 12th.

A really warm day.

Miss Erskine (very handsome), and Admiral

---

* This is the same Mr. MacCallum whom we met at Rose Bank, and who brought the Persian Greyhound. (*See* Journal, June 9th).

† The remarkable effect of light is the glare of the bombardment contrasted with the light of the moon.

1876. Spencer came to luncheon.   Went out driving with
Fanny and Rose; we saw Lady Bell and Katharine.

Tuesday, 13th.

Leonard Lyell and his wife, and Nellie, Effie, and
Blanche Sanford came to luncheon.

Harry arrived. Minnie and Clement dined with us.

Wednesday, 14th.

Went with Fanny and Rose to Veitch's nursery
garden, and spent an hour-and-a-half there very
pleasantly.   It is a wonderfully rich collection of
beautiful and curious plants.   Rose and I recognised
with great pleasure several of our old tropical
friends, with which we had made acquaintance, she
in the W. Indies and I in Brazil.   There were also
very attractive houses full of lovely Australian
and Cape plants; Tetrathecus, Boronias, Draco-
phyllums, Darwinias, Heaths, Helichrysums, &c.
(but no Proteas).   The vast number of plants
of several different species of Nepenthes was quite
surprising, and the size and beauty and various
forms of their pitchers very interesting; some really
superb.   Scarcely anything interested me more than
Cypripedium spectabile, of which there were several
plants (almost in the open air), flowering beautifully,
and apparently in high health; the flower white,
with the lips exquisitely flushed with pink.   Among
the hothouse Orchids was another lovely Cypripedium—niveum—one of the Indian group with no
leafy stem ; a small but delicately formed flower of
an exquisitely pure white.   Maregravia, a curious

and botanically-interesting plant which I never 1876. before saw cultivated ; not in flower, but showing some of the curious variability of its leaves on the creeping stem, and on the spreading branches. Tabernæmontana coronaria, in flower, very beautiful and fragrant.

Bertolonia; several beautiful varieties or supposed species, with curious variations in arrangement of spots on the leaves.

"Ampelopsis Veitchii" (properly Vitis), growing very luxuriantly over some low walls in the open air (they say here it is quite hardy), quite as rich and luxuriant in its growth as the Virginian creeper; the leaves curiously variable, compound (of three or five leaflets) and simply lobed ones, mixed together.

Fanny and I dined with the Bishop of Winchester —met the Bishop of Gloucester and Derry, Canon and Mrs. Barry, Canon Prothero, Lord Crewe.

———

Thursday, 15th.

Our dinner party : the Sanfords (including Nellie who looked remarkably pretty), Lady Lilford and her niece Miss Brandling, Lord Napier, the Mac-Murdos, the Lynedoch Gardiners, Admiral and Lady Sotheby, Mr. and Mrs. Drummond, Minnie, Harry Bruce, Edward, and our nephew Harry.— Very pleasant. I was between Lady Lilford and Lady Sotheby, both agreeable.

———

Friday, 16th.

We dined with the Locke Kings : met Sir Bartle

1876. and Lady Frere, Mr. Massey, Colonel and Lady
Elizabeth Romilly, Mr. and Mrs. Thomson Hankey,
Mr. Kinglake and others. I renewed my acquain-
tance with Kinglake, whom I had known long ago at
Cambridge, but had hardly met since. I am glad
to know him again.

Saturday 17th.

Isabel Hervey dined with us, and she, Rose,
Minnie and Harry went to the Opera. Fanny and
I stayed at home.

A visit from Cecil.

Again to the Royal Academy, but only a hasty
visit.

Another startling piece of news from Constantinople—the assassination of two of the Turkish
Ministers, while actually sitting in Council (and of
some other men), by a Circassian officer. Turkey
has certainly supplied us of late with an extraordinary quantity of " sensational " intelligence.

Monday, 19th.

A really fine warm day.

Went to the Athenæum and voted for Mr.
Oliphant.

A very pleasant visit from Carrick Moore.

Our guests at dinner:—besides Minnie Napier
and Harry Bunbury, who dine constantly with us,
—Sarah and Albert Seymour, Lord and Lady
Rayleigh and Miss Alice Balfour, Sir Francis and

Lady Boileau, Col. and Mrs. Legge, Willoughby 1876.
Burrell, Mr. and Mrs. Walrond, Matthew Arnold,
Mr. Bouverie.—A pleasant party. Many more (a
" drum ") in the evening.

Tuesday, 20th.

Very fine and warm.

The Sanfords came to luncheon with us, and we
went with them to Mr. Mac Callum's at Kensington,
to see his landscape paintings. Unfortunately he
had been summoned away. His pictures appeared
to me to be of great merit ; several views in Egypt,
in particular, with the marvellous glow on the ruins
and the bare mountains, very beautiful and full of
character : also views near Cannes, and some
admirable groups of old fir trees from the forests
of the Scottish Highlands. Mr. Sanford admired
them very much.

Wednesday, 21st

Very fine and warm.

Went to the Linnean Society and got the vols. of
the Transactions due to me.

Mr. and Miss Galton, May Egerton, Alfred
Newton and Edward dined with us—very pleasant.

Thursday, 22nd.

Admiral Spencer came to luncheon, and went
with us to the MacMurdos' afternoon party at

1876. Rose Bank.   Fanny and Rose went to Kate
Hoare's "drum."

Our guests at luncheon :—Lady Harriet Hervey,
Mrs. Mills, Frank Lyell.

Our dinner party (in addition to our regular daily
party) : the Charles Hoares, Katharine and
Rosamond, the Maskelynes, the Charles Mallets,
the Arthnr Milmans, the Coltmans, Mr. Bentham,
young Mr. Locke King, Clement.—A very pleasant
party.

I sat between Kate Hoare and Mrs. Maskelyne, whom I scarcely knew before, and whom
I found exceedingly agreeable and interesting.
She is a zealous botanist, and seemed to take
delight in talking on the subject, so that we
sympathized heartily.

Mrs. Maskelyne is a granddaughter of Lewis
Weston Dillwyn, the botanist, and she says there
have been three generations of botanists in her
family.

Her interesting account of Mr. Moggridge—
who was a first-cousin of hers—how, for fourteen
years, he was in such danger of consumption, that
he was obliged to spend much of every year at
Mentone ; how his devotion to natural history kept
him always cheerful and happy in the midst of all
his sufferings from ill-health, and when he appeared
on the grave, and how he cheered many fellow
sufferers by interesting them in the same pursuits,

and showing them the natural curiosities which he 1876. had observed.

—————

Lord Napier came to luncheon—very agreeable. Talked, *inter alia*, of Mr. Mark Napier's inquiries into the history of Graham of Claverhouse, and the new light thrown on his character—that Claverhouse was not more cruel than most men of his time, and not at all a romantic character, but a brilliant soldier and a shrewd, sagacious, worldly man, with a keen eye to his own interest.

We (Fanny, Rose and I) went to the Botanic Society's Gardens in the Regents Park, and enjoyed a stroll in them.—(I have lately become a Fellow). The great conservatory afforded a beautiful sight— many fine specimens of palms, bananas, ferns, tropical aroids and climbers, and other noble forms of vegetation.

Fanny and I dined with the Aberdares—a pleasant party:—Lord and Lady Leigh, the Lynedoch Gardiners, Mr. and Miss Childers, a Professor from Cambridge, Massachusetts, with his wife (the name Gurney or something like it), Mr. Aubrey de Vere, a Spanish savant whose name I did not catch, and one or two more. I sat by the American lady, whom I found conversible, intelligent and agreeable.

—————

Very fine.
We went to see the Loan Exhibition of Scientific

1876. Apparatus at South Kensington. The greater part of it was bewildering and unintelligible to me for want of previous knowledge; the distances are enormous, and we were fatigued before we reached the departments (those of geography and geology) which were interesting to us. We saw, however, Galileo's instruments, Newton's telescope, and some made and used by Sir William Herschel, and other objects of historical interest. In the geological department, the beautiful fossil plants from the tertiaries of Austria, arranged by Ettingshausen, are especially interesting.

Miss Froude, daughter of the great historian, came to luncheon. She is a very pleasing, attractive, interesting girl, and evidently very clever. She looks at first sight like a pretty child, but her intellect is far from childish.

Our dinner party :—Lady Mary Egerton, the Sanfords, the Locke Kings, Lady Head, Mr. Kinglake, Lady Wood and Mr. Cox, Mr. George Loch, Mr. Eddis, the MacMurdos, Clement.

Poor Kinglake looks wretchedly ill.

————

Wednesday, 28th.

Fanny went with MacMurdo to Haileybury, to see Arthur. Rose spent the day with Sarah Sanford. Minnie and I went to Kew Gardens, and spent the afternoon very pleasantly :—went through the tropical and temperate fern-houses and the great palm-house. The ferns in wonderful beauty and variety, very healthy and flourishing :—

noble tree-ferns and grand plants of Angiopteris, the fronds of which are almost the largest of all, though it is not arborescent,—beautiful Lygodiums of several species, a number of exquisite Trichomanes and Hymenophyllums in glass cases, looking very healthy and glittering with moisture. In the palm house, the multitude of grand specimens of stately palms, Pandani, and some other exotic trees, make a magnificent show, and there are great numbers of other interesting plants of smaller size.

A great number and variety of small and delicate tropical ferns cultivated here on dead trunks of tree ferns, just as they grow on living ones in the tropical forests—as I have seen them in Brazil. A great variety of Bromeliaceæ, more than I have ever before seen in cultivation, reminding me strongly of the Brazilian forests, in which plants of that family are so abundant.

The greater part of the day was beautiful, but while we were in the palm-house a most sudden and violent thunder shower came on, just like a tropical shower, but we did not hear any thunder. It did not last long.

———

Thursday, 29th.

We went to a garden party at the Archbishop's, at Lambeth: it was very numerous, and we met many friends and acquaintances.

———

Friday, 30th.

Our guests at luncheon: — Katharine and

1876. Rosamond, the MacMurdos, Mrs. T. Thornhill, Mr. Charles Murray and Mr. Sholto Murray. Katharine told me that Joseph Hooker is engaged to be married to Lady Jardine, the Widow of Sir William, and daughter of Mr. Symonds, the geologist. I heartily wish him happiness, for he is an excellent man, as well as a most able one.

Our dinner party :—Lady Octavia Legge, the Albert Seymours, the Sanfords, Lady Head and her daughter Amabel, Pamela and Philip Miles, Major and Mr. Grant, Edward, Admiral Spencer, Mr. Harrison, Mr. Fletcher.

This was an exceedingly pleasant party. Sarah Seymour lovely—even more so than usual. Sarah Sanford in good looks and spirits ; it is delightful to see her so much restored to her former self, no longer looking so painfully fragile and ghost-like.

———

Saturday, July 1st.

Dear Rose Kingsley went home. I was very sorry to part with her. She has been a delightful inmate of our house for seven weeks and more.

Mrs. Galton and her daughter came to luncheon. Drove out with Minnie and Fanny : visited the Zoological Gardens.

———

Monday, 3rd.

We said farewell to dear Minnie and Sarah, and returned to Barton.—All well at home—thank God.

———

Tuesday, 4th.

Quarter sessions at Bury ; address approving of the Prisons Bill, proposed by Phillips, seconded by me, and voted without opposition.

Clement and another young barrister, Mr. De Grey (a very pleasing young man) dined with us.

Wednesday, 5th.

We drove out in the pony-carriage. Looked at the home farm with Scott.

Thursday, 6th.

Strolled and looked at the haymaking and the trees, &c.

Friday, 7th.

We drove to Stowlangtoft ; had a long and pleasant talk with Mrs. Wilson and Agnes ; met Lady Hoste as we returned.

Saturday, 8th.

Fanny went to Mildenhall and returned in good time.

Sunday, 9th.

We went to morning Church, and heard an excellent sermon from Mr. Percy Smith on Philemon.

Weather beautiful ever since we returned home, and Barton at its best—thoroughly enjoyable. Hay

1876. harvest going on admirably; prospects for the harvest very promising. Foliage of the trees very rich and luxuriant. The Indian Horse-chesnut— Esculus (Pavia) Indica, flowering beautifully, more abundantly than I remember ever to have seen it before. The Robinia viscosa (which is an uncertain flowerer) is flowering well, while the blossom of the common white Robinia is not yet past. So also the flowers of the common Rhododendron still linger on. Three of our trees of the Indian Horse-chesnut are now in flower, one of them (" Emily's tree ") for the first time.

We have heard the Cuckoo frequently since we came home. I think it is late for this.

On the 5th I wrote to dear Mrs. Kingsley, from whom we had jointly received a delightful letter, overflowing as usual with love and kindness to us both.

On our return home I found among other letters awaiting me, one from the foreign secretary of the Royal Society, asking for information about Adolphe Brongniart, and for my opinion about his scientific character. I had no information to give about his personal history, but I wrote yesterday, a careful letter of remarks on his scientific character and merits.

———————

Monday, 10th.

Mr. Percy Smith came to early dinner with us. Agnes Wilson and Mrs. Byron later in afternoon.

————

Gave up journey to Stutton on account of Mr. Mills's illness.

Ramble with Fanny and Scott.

We returned the day before yesterday, from a two days' visit to the Bernerses at Woolverstone. It was very pleasant. I have already repeatedly described in my journal, that beautiful place—the most beautiful to my thinking, in Suffolk, with its varied surface, and fine, bold slopes down to the river ; its noble trees, its " Fernery on the Cliff," (a botanical Paradise), and its rich gardens. All are now in full beauty. The weather was splendid and very hot.

Mr. and Mrs Berners appeared in good health, and as we have always found them, full of good nature, cheerfulness, and good-humoured friendliness.

In the beautiful conservatory at Woolverstone, I saw three things which struck me as remarkable :— Strelitzia Reginæ in flower, in the Fernery of the conservatory, in a very moderate temperature, by no means tropical. Lapageria rosea, bearing *fruit* (unripe). Clerodendron Balfourianum with ripe and open fruit, showing the seeds with their bright orange-red arils.

We were the only guests at Woolverstone, except the clergyman, Mr. Wood.

On the 14th, the Bernerses took us over to

1876. Stutton to see the Millses. Mrs. Mills seemed very well, and quite unchanged. Mr. Mills had had an attack of bronchitis, and was confined to his room, but we were allowed to see him for a few minutes, and he was as friendly and affectionate as possible.

There was a garden party at Stutton, where we were glad to meet Lady Florence Barnardistone and her sister, Lady Louisa, that exceedingly pretty woman Mrs. Robert Anstruther, and a few others.

George Hervey was married on the 13th, to Miss Arkwright, and Fanny had a very pleasant letter, yesterday or the day before from Sarah, giving a lively account of the wedding. There was a large family gathering. The marriage is very satisfactory to our dear friends.

The hay-harvest in our park is pretty well finished, and the weather has been superb for it all the time.

———

Tuesday, 18th.

My near wife's 62nd birthday. My first acquaintance with little Charlie Lyell.

A garden party at the Victor Paley's.

———

Thursday 20th.

Lady Florence Barnardistone and Lady Louisa Legge arrived. Lady Hoste dined with us.

Scott tells me that, in his experience of 35 years on this estate, he does not remember one occasion on which the season of the hay-harvest was more entirely favourable than it has been this year.

There has been no rain at all from the beginning to 1876. the end of the time.

Weather for these three weeks constantly fine, clear and dry,—latterly very hot.

Friday, 21st.

Agnes Wilson and her brother Arthur and Mr. Byron, dined with us

----

Saturday, 22nd.

Lady Louisa and Lady Florence left us after luncheon.

Captain Horton dined with us.

Fanny had a charming letter from Rose Kingsley, giving a most agreeable account of her sister Mary's wedding with Mr. Harrison, which took place on the 18th. In every circumstance, it seems to have been everything that could be desired, and, most happily, Mrs. Kingsley was (for her) remarkably well, and does not seem as yet to have suffered from the fatigue and excitement.

Monday, 24th.

Some heavy showers in morning, and air much fresher than before.

Harry and George arrived. Captain and Mrs. Corry, Mr. J. Bevan and his daughter Evelyn dined with us.

Thursday, 27th.

Dear Arthur arrived.

LETTERS.

1876.   My Dear Katharine,

Very many thanks for your interesting
letter from Pendock, and for sending me the
proposed inscription for Charles Lyell's tomb, which
(the inscription) I entirely approve. Indeed, what
was drawn up by Dean Stanley was sure to be
good. It expresses very well, I think, Charles
Lyell's peculiar and characteristic merits as a
philosopher, and the great work of his life. I join
with you in hoping that his bust may one day be
set up in the Abbey.

I am very glad you are spending a pleasant time
at Pendock, which I daresay is a very agreeable
place in fine weather. When I saw it was in the
depth of a particularly wet autumn, and the lanes
were deep in red mud ; but the view of the
Malvern Hills from that plain is always beautiful.

I heartily wish happiness to Joseph Hooker and
his bride that is to be ; I have a very high opinion
of him, a great esteem for his character, as well as a
great admiration of his scientific works.

Your little grandson (and my godson), is a dear
little fellow, very pretty, very amiable and good-
humoured, (I have never yet heard him cry !) and
seems very intelligent for his age.

Mary is extremely pleasant ; I like her more and
more, and think more highly of her, the better I

know her. I am very glad to have her company 1876. here, and I hope she likes Barton.

I was sorry that Leonard was called away so suddenly from us, to make such a long and fatiguing journey to Scotland in such hot weather, but I hope he will be amply compensated for it.

The new number of *The Quarterly* seems indeed, as you say, to contain many good articles ; I have not yet read *thoroughly* more than two of them : on Mr. Ticknor (very entertaining), and on Trees (excellent). There are very good articles in *The Edinburgh* also ; those I have read are :—on the Comte de Paris' Campaigns in America (during the Civil War) : on Van Ranke's History of England, and on Moresby's New Guinea.

I am going carefully through Wallace on the " Geographical Distribution of Animals," which is a wonderful collection of curious and well-arranged information,—such a mine of knowledge, that it will require reading more than once. I am also studying the new volume of Hooker and Bentham's " Genera Plantarum," which must have been a work of immense labour. But in truth, I have more new books than I can read. I have lately got Mr. Moggridge's " Flora of Mentone," a beautiful book, which would be very useful if I ever went to the Riviera again.

Fanny is tolerably well, and sends you much love. We have glorious weather, and have had the finest hay-crop that has been known for many years.

Ever your very affectionate brother.
CHARLES J. F. BUNBURY.

1876.   P.S.—I owe you hearty thanks for the beautiful fender-stool which you were so good as to work for me, and which is now a great ornament to my hearth.

Barton Hall, Bury St. Edmund's,
July 28th, '76.

My Dear Katharine,

I had finished my letter to you, and put it in the box, before Mary received her letter from Leonard with the good news about Mr. Mudie's will, which has quite delighted us, and I cannot help writing this line to congratulate you most heartily. I have not heard such good news for some time. I honour the old gentleman for disposing of his wealth so well and wisely. I hope Leonard's share will prove, in the long run, as good as it promises; yours, and your other children's are most pleasant and satisfactory to hear of, and I am sure that wealth could not come to people better qualified to make a good use of it. I feel heartily for and with Mary.

Ever your affectionate brother,

CHARLES J. F. BUNBURY.

JOURNAL.

Friday, 28th.

Edward and Clement arrived. George went away.

Saturday, 29th.

Lady Hoste came to luncheon. Agnes Wilson and a party of children later in afternoon: the children's games.

———

Monday, 31st.

Annie Campbell, with her little sister Finetta, and four of their brothers, are with us now, having arrived on the 28th. Arthur MacMurdo came the day before. Annie (aged 17) has a lovely face, a cultivated mind, a charming disposition and character.

———

August 1st (Tuesday).

The Assizes at Bury. I foreman of Grand Jury; a hardish day's work.

Dined with the Judges, Chief Justice Cockburn and Baron Huddlestone.

———

Friday, 4th.

Harry, Clement, and Mr. De Grey went away.

The Rifle meeting in Ickworth park—distribution of prizes—speeches.

———

Saturday, 5th.

All the Campbells went away early, and Arthur with them.

———

Monday, 7th.

Leonard and Mary Lyell and their baby left us this morning, after we had planted a tree (an

1876. Indian Horse-chesnut, Æsculus, or Pavia Indica),
in commemoration of their visit.   Leonard returned
from Scotland on the 4th.   Everything relating to
Mr. Mudie's bequest, and especially to Leonard's
fortune, is most satisfactory—delightfully so.   A
day or two ago Fanny had a letter from dear Rose
Kingsley, written in a tone of great happiness.   Her
brother Maurice and his wife have arrived from
Ameriea, and are now with them at Byfleet, which
makes them all, and especially Mrs. Kingsley, very
happy.

_____

Wednesday, 9th.

Up to London.   Minnie and Arthur met us at
48, E. P.

_____

Thursday, 10th.

A visit from the MacMurdos.   Down to Folkestone (Arthur with us), to Bates's hotel.
Cissy and Emmy came to us in evening.

_____

Friday, 11th.

Folkestone.   Most beautiful weather.   Walk in
morning along the cliff towards Sandgate.
We drove with Cissy and Emmy, Edward and
Frank Lyell, to Hythe, saw the church, &c.   Cissy,
Emmy, Edward, F. Lyell and Guy Campbell dined
with us.

_____

Saturday, 12th.

The same brilliant weather.   Morning stroll on

the cliff. Conversation with Frank Lyell. Edward 1876. set out for France. We visited Frank in his hut at Shorncliff Camp.

———

Weather excessively hot. Read prayers with Fanny, Arthur, and Susan Kerry.

———

Again excessively hot. We had a pleasant drive with Cissy and Emmy through pretty lanes by Cheriton, Saltwood and Hythe. Guy Campbell and Willie dined with us.

———

We went down to Folkestone on the 10th, to Bates's hotel, and returned yesterday. Our object in going thither was to see dear Cissy and Emmie, who are staying there, and have been for some time, for the benefit of Emmie's health. Accordingly we spent the greater part of each day in their company.

The weather has been excessively hot, bright and dry, to an extraordinary degree indeed. All this last week, except only the 14th, when there was a thick fog from the sea, and unluckily this was the day we had fixed for a visit to Dover.

The 14th we drove to Dover, we two and Cissy in an open carriage ; Arthur having gone by railway in the morning, met us at the *Lord Warden*. Drive about 1½ hour each way. Long ascent from Folke-

1876. stone to the top of a very high Down, on that side
mostly uncultivated and grassy ; abundance of
pretty wild flowers, even in this parching season,
among the grass, but the Origanum, which was
in great plenty, the only one I noticed we have not
at Mildenhall. Descent on the Dover side through
a long, dry valley, the open hill sides which slope up
on either hand in great part cultivated. Very large,
open cornfields without fences : very little wood in
sight.

At Dover we went up to the Castle, and saw the
curious old Church, and the Roman tower (originally
a Pharos) ; but the persistent fog prevented us from
seeing any view. Much impressed however, by the
grandeur of the situation of the castle and the
massive air of strength.

———

Wednesday, 16th.

We returned to London. Continued hot and
dry weather. Minnie dined with us.

———

Thursday, 17th.

Excessively hot.

Minnie, Sarah and Albert and Susan MacMurdo
dined with us. Very pleasant.

———

Friday, 18th.

Down to Wells—delighted to be there.

———

Saturday, 19th.

In the delightful Palace at Wells, where in '70

and 71, we spent such happy days. Received by 1876. our dear old friends, with all their old kindness and cordiality. The family party however is much reduced, of course, by the absence of the three married daughters, and it seems almost strange to find the Bishop and Lady Arthur with only one daughter (the youngest) with them and two sons, Sydenham and James.

Sunday, 20th.

We went with the Arthur Herveys to morning service in the Cathedral.

Thunder and rain in afternoon.

Monday, 21st.

We visited the Dean and Mrs. Johnson.

The Bishop took us in his light open carriage to one of these charming spots on the slopes of the Mendips, of which he has found so many. The party were himself, Fanny and I and Caroline or "*Truey*" Hervey, the only daughter remaining unmarried,—a charming girl. Leaving the carriage in a narrow and rugged lane, we drank tea very merrily on the turf, and then had a delightful walk up the smooth grassy slope to the broad back or table-land, commanding a glorious view. The day was beautiful, the changing clouds very picturesque, but the distance hazy ; the Blackdown hills in-distinct, and the Quantocks hardly visible at all. A sudden gleam of sunshine, from amidst the

1876. clouds, lighted up the sea (Bridgewater Bay) with great brilliancy, and the singular bold hill, Brent Knoll, stood out boldly against it.

The view was in the main the same as those which I described several times in '70 and '71, seen from neighbouring though not exactly the same spots. It is delightful scenery, from the richness, greenness and luxuriant variety of the plains and nearer lower hills, and the variety of ranges and heights which rise at successive distances. The several underfalls of the Mendips, soft round green hills, diversified with wood and pasture, form very agreeable features in the foreground. One of these which was directly below us to-day, is Milton Hill, to which I remember I walked with Sarah and Kate in '71.

Mr. and Mrs. and Miss Gambier Parry, Mr. and Mrs. Neville Grenville and others came to the Palace.

———

Tuesday, 22nd.

Expedition to Cheddar.

Mr. Freeman, the Historian, and Professor Babington came to dinner.

———

Wednesday, 23rd.

The party to Cheddar yesterday, under the Bishop's direction, was a numerous one:—ourselves, Caroline Hervey, the Gambier Parry's, the two Miss Nevilles, Miss Arkwright and her brother,

Hugh Hoare and James Hervey. The day was 1876.
fine and the party pleasant.

————

We left Wells and went (by Glastonbury and
Temple Combe) to Salisbury.

Most kindly received by Mr. and Lady Barbara
Yeatman. They have a pretty house with an
excellent garden in the Close. Lady Barbara was
one of the Legges. Mr. Yeatman is a remarkably
agreeable and well-informed man. His sister, Miss
Yeatman, who was staying with them, very pleasant
and very intelligent.

We went with them to the Cathedral, which was
swarming with archæologists (for the county Archæological Society was holding its meeting at
Salisbury). The west front has been beautifully
restored. The choir closed for repairs.

In the evening an archæological meeting in the
Blackmore Museum. Papers read:— 1. By Mr.
Maskelyne, on the Petrology (the mineral composition) of Stonehenge ; elaborate and curious. 2.—
By Rev. Mr. Morres, on some rare birds lately
observed in Wilts. 3.—By Canon Jackson on the
true history of Amy Robsart.

Tecoma (or Bignonia) radicans—a fine plant,
flowering beautifully, against Mr. Yeatman's house
at Salisbury.

————

A long and fatiguing but interesting day.
Weather very fine and bright, but wind rough

1876. and cold.   Arthur started with Mr. and Miss
Yeatman and the archæologists at 9; we by ourselves in an open carriage at 11.   We went first to
Old Sarum, and spent some time very pleasantly
in rambling over it.   It is extremely interesting.
Situation very striking.   It is a sort of promontory
of the chalk downs, rising boldly from the plain on
three sides, and commanding a wide view of the
low grounds to the opposite hills.   The great trench
which formed the line of defence of the city is well
preserved, enclosing an elevated plain, which was
the site of the city, cathedral, &c. (now occupied by
cultivated fields).   One massy fragment of masonry
remains, but of uncertain date.   Citadel, rising
high above the site of the city : its trench very
grand, cut (like the outer one), in the solid chalk,
with sides all but precipitous, now clothed with
thickets.   The magnitude of these trenches, cut in
the solid chalk is very striking.   Some fragments of
masonry by the side of the path by which one
ascends to it.   Good views of Old Sarum from
several points in this day's drive ; it stands out
against the sky like an actual castle.

From Old Sarum, by Lake House to Amesbury,
along the pleasant green valley of the Avon,
between two promontories of the chalk downs, the
river smooth and still and gently winding.

At Amesbury we did not see much, but not far
from it, Mr. Yeatman, whom we had now rejoined,
guided us to the so-called *Vespasian's Camp*, which
is interesting.   It is an ancient entrenchment,
(Roman or perhaps originally British), on the side

of a very high grassy bank (or rather indeed a very 1876. steep hill-side), facing and overlooking Sir Edmund Antrobus's fine large new house. This house beautifully situated—a broad green gently-sloping lawn in front, half encircled by the quiet stream of the Avon, which skirts the foot of the entrenched height —woods behind the house and woods also on the height flanking the entrenchment.

Thence to Stonehenge.

Here the Archæologists mustered in force, and there was much discussion and great differences of opinion—harangues from Mr. Parker, Sir John Lubbock and others.

We ended with Bemerton, a village very near to Salisbury, on the bank of the Wiley, a small river which joins the Avon at that city. Here are the church and parsonage of the religious poet George Herbert; the church very small; the parsonage (kept with great care by the present incumbent), very pretty, quiet, and gentle-looking, with its pleasant garden sloping down to the river.

—— — ——

Friday, 25th.

Salisbury.—Visited the Blackmore museum. The collection of pre-historic antiquities is very remarkable and instructive. A large collection of stuffed birds, which would be more interesting if they were sufficiently labelled. The most worth notice appear to be two Bustards (? both female), killed near Salisbury in 1871, the last killed (or seen, I believe) in England.

1876.   The weather still fine, but had now turned very
cold. We drove with Mr. and Miss Yeatman to
see what is called *the Moot ;* a puzzling piece of
antiquity. It consists of several earthworks, of considerable size and apparent importance, but of
which I could not understand the plan or purpose.
They are said to be Saxon.

Saturday 26th.

From Salisbury to Sherborne.   The Digby Hotel
at Sherborne very good.

The church is very large, cruciform, with a great
square tower in the centre ; and standing in the
highest part of the town, has a very fine and
dignified appearance, especially as seen from the
direction of the railway station.

We dined with Mr. and Mrs. Hutchings at
Sandford Orcas, four miles from Sherborne ; met
Mr. and Mrs. Talbot Baker and Mr. Portman.

Sunday, 27th.

We attended morning service in the beautiful
Abbey Church, which is quite like a cathedral, and
looks as if it ought to be one.   After the service we
went more carefully round the interior of the
church. It is very beautiful indeed. It was originally
Norman, and the Norman character remains visible
in some arches and some details of ornament here
and there ; but generally the Norman has been
replaced by Perpendicular of the richest style.

Fan tracery of the roof, especially beautiful. The 1876. whole has been most superbly restored by Mr. Digby.

In the afternoon we drove through Mr. Digby's park, which begins at the very outskirts of the town, and is fine and very extensive; considerable variety of surface; abundance of fine timber (in particular some very grave old oaks); plenty of fallow deer (many of the dark brown variety). We passed near Sherborne Castle, the present house, which was the property of Sir Walter Raleigh, and inhabited by him. In another part of the park are the fine picturesque, massy ruins of a much older (Norman) castle, which was taken by Fairfax in 1645, and destroyed by order of the Parliament.

Thence by a long circuitous way which showed us much of the country, to Sandford Orcas, to drink tea with the Hutchings. Country hilly, varied and pretty; much grass land and hedge-row timber; very deep lanes and hollow ways, in which the yellowish-brown hue of the oolite is everywhere conspicuous, and the strata in many places well displayed.

Clematis vitalba, Origanum vulgare and Viburnum Lantana,—prevailing plants on the oolite about Sherborne, as well as on the chalk about Folkestone and Dover.

———

Monday, 28th.

From Sherborne by road to Yeovil. Then by railway to Taunton. Thence to Minehead, also by railway.

1876.   Quit the oolite country between Yeovil and
Taunton, and enter on the low alluvial country.
Pass Langport, where Fairfax defeated the Cavaliers
in 1645.

Thence to Taunton, a long stretch of low marshy
meadows, in which we pass Athelney, where Alfred
defended himself against the Danes.   As we passed
over this marshy country, Arthur observed more
than one hen-harrier.

Beyond Taunton the Blackdown Hills in sight on
our left, (to South), the Quantocks on our right :
the intermediate low country (the Vale of Taunton)
beautifully rich and verdant : the formation, New
Red or Trias, as is very evident by the colour of the
broken banks and ploughed fields.   The Quantocks,
at first faint and distant, but as we approach
them, growing into dark swelling heathery masses,
the railway going partly round their western
extremity.   Touch the sea-shore at Watchet, where
conspicuous masses of lias.

Minehead, a long stretch of low flat shore
between two bold headlands.   Myrtles and
magnolias growing luxuriantly, and flowering
against the walls of the houses.

Tuesday, 29th.

From Minehead to Lynton, 4½ hours by road
(with two carriages and six horses in all), over the
enormous hills, part of Exmoor.   From Porlock, a
beautiful new road takes us up to the brow of the
heights, through pleasant young woods, the road

much less steep than the old one, winding 1876. judiciously, delightful views opening at every turn between the trees. Having ascended thus to the brow, we had next a long drive over the bare, bleak, heathy hills, well beaten by the wind, for though the day was fine, the wind was very strong and rough. Elevation of this part of the road (according to Murray) about 1,100 feet. Thence a very long and steep descent, winding along the open seaward face of the hills, to the beautiful nook in which lies Lynmouth, and lastly, the excessively steep ascent to Lynton.

Wednesday, 30th.

Valley of Rocks Hotel, Lynton.

Violent storms of wind and rain in afternoon, but we went to Woody Bay and spent the afternoon very pleasantly with the Sanfords, Sarah, Mr. Sanford, the four girls and two boys. Mrs. and Miss Bouverie staying with them.

Thursday, 31st.

Weather furiously bad all morning—in afternoon better, though blowing hard.

Mr. Murray and Harry had luncheon with us.

We drove to Woody Bay, and drank tea with the Sanfords.

September 4th.

Lynton.

The extremely bad weather, every day except the

1876. 1st and 2nd, rather interfered with our enjoyment of this beautiful place.   It was the fourth time I had seen it, and I certainly did not think its beauty overrated.

We spent many delightful hours with the San-fords, seeing them every day but one during our our stay.   Their home at Woody Bay, a charming spot, is about 4 miles W. of Lynton; the way to it, first through the Valley of Rocks and past the house of Lee, then, after some steep ups and downs, along the seaward face of the hills, winding through Mr. Sanford's beautiful woods, with the steep hill-side above, and the still deeper descent to the actual cliff below.—A very beautiful drive, but rather perilous looking.   The road very narrow, with no sort of parapet or protection, and the descent on the outer side so nearly precipitous, that one wonders how trees can grow on it.   In many places, as one looks over the edge, the view plunges directly down to the sea, which seems to be directly below.   This drive reminds me somewhat of "the Hobby" at Clovelly, but the woods are less rich and the trees not so large.   The undergrowth of heath, ferns, wood-rush and mosses, very beautiful and luxuriant here.   Descent at last to the house by several sharp turns.   The house—small and pretty—stands embosomed in wood, on a very small level space, perhaps 200 feet above the sea, and probably not much less than 1,000 feet below the tops of the hills.   Immediately below it is a small cove or recess among the rocks, where (at least at low water and in calm weather), a little

beach is left between the sea and the cliffs. On 1876.
the further side of this cove, directly facing the
windows of the house, is a magnificent cliff,
absolutely precipitous, of bare rock, the more
striking from its deep fiery red colour. Above,
and alongside of this precipice, the cliff is richly
clothed with dense woods, extending far up the
hill side, and beyond this nearest cliff, looking
along the coast towards Lynton, we see one
picturesque craggy headland after another, with
the sea foaming on the dark rocks at their feet.
The wretched weather prevented us from exploring
the rocks and woods even near Mr. Sanford's
house.

Mr. Sanford's family were all at Woody Bay at
this time : four daughters—Ellen (or Nellie), Ethel,
Blanche and Rose—and two sons, Ayshford and
Harry, also Mrs. Bouverie (sister of the first Mrs.
Sanford), and her daughter. The Sanford girls are
charming mountain nymphs—frank, joyous, full of
good nature and high spirits, and intelligent
besides : the sons, fine specimens of boys. Sarah,
I am sorry to say, was not looking well.

Some of the most conspicuously common plants
in these woods (and I think in North Devon
generally), at this season, are Solidago, a large
Hieracium (probably a broad-leaved variety of
umbellatum), Luzula sylvatica (growing very large),
Melampyrum pratense (always with deep yellow
flowers). The ferns that I saw were common
kinds, the least general of them being Nephr.
Oreopteris ; but Mr. Sanford says that the

1876. Hymenophyllum grows beside a waterfall on his property. The Pyrus Aria, usually confined to limestone or chalk, grows in abundance (as Mr. Sanford pointed out to me) on the cliffs at Woody Bay (and I saw it also in the valley of the East Lyn); Pyrus torminalis, he said, also grows in these woods, but more scattered.

On the 1st September, young Sanford brought into Woody Bay, besides partridges, two brace of heath-poults (young blackcocks), which he had shot on the moors above. The plumage is almost exactly the same as that of the hen bird.

Mr. Sanford tells me that the Peregrine falcon and the Cornish clough make their nests in the cliffs near Woody Bay. The buzzard is common on the moors in this part of the country: the hen harrier rare.

On the 2nd September the weather was better, and we were able to join the Sanfords in an excursion up the valley of the East Lyn: it was very agreeable. We were a large party:—all the Sanfords (including Mrs. Bouverie), my nephew Harry and Mr. Murray and Lord Tenterden (who has a pretty cottage in a lovely spot in the lower part of the valley), likewise joined the party. Fanny, Sarah and Mrs. Bouverie sat down aud sketched : I, with Mr. Sanford, his daughters and Lord Tenterden walked up the valley, by the foot-path along the margin of the stream, to " Waters Meet," I had very great pleasure in seeing again this beautiful valley, with which I used to be so familiar 45 years ago. It charmed me as much as ever. The combinations of wood, water, rock and 1876. hill are delightful. Lord Tenterden (justly, I think) compared this valley to the Trosachs, and to the Dargle in county Wicklow. With the valley of Rocks we became very familiar, as we drove through it every day that we went to Woody Bay, and moreover, spent part of two afternoons there, Fanny sketching, and I hunting for Lichens. It is a very fine and interesting piece of rock scenery.

The bad weather which hampered our operations on land, at least gave us the advantage of seeing the sea in a great variety of picturesque aspects. At times, especially in the mornings, we saw the opposite coast of Wales (Glamorganshire) with beautiful distinctness; the high hills of the Coal district rising obscure, but dark and massy, in the back-ground beyond the immediate coast.

Mr. Sanford finds a great deal of occupation in collecting and studying the marine animals of Woody Bay. His dressing room was crowded with microscopes, collecting apparatus, basins and tubs and jars full of salt water and the like. He showed me some very beautiful microscopic objects which he had prepared: in particular, I remember, the tongue of a limpet, which, under a high magnifying power, looked like a band covered with exquisite jewellery.

———

Tuesday, 5th.

Left Lynton with regret.

From Lynton to Barnstaple, by road (with two

1876. carriages and six horses), three hours, over a succession of very high and steep hills. First, a long ascent up the very beautiful and richly wooded valley of the West Lyn : then some miles over a high wild open moor, now exceedingly gay with heath and furze in full blossom : then a long descent to the village of Parracombe, situated in a deep hollow. Thence a succession of ups and downs to Barnstaple, the road in many parts very steep, the hill sides often richly wooded, bright green meadows occupying the bottoms of the valleys. A remarkable profusion of Ferns all along the road-sides on this route, clothing the bank and low walls with a most luxuriant and beautiful growth.

From Barnstaple to Taunton by railway, a little more than two hours, passing by South Molton, Dulverton, Wiveliscombe and Milverton. A beautiful rich country, verdant, cheerful and varied. Near the Castle Hill Station, not far from Barnstaple, we cross by a viaduct a part of Lord Fortescue's beautiful park. *Taunton*—a large church, with a very tall and very fine tower of the Perpendicular style (rebuilt in 1857-62), the material, sandstone of a very peculiar purplish red or almost pink tint.

————

Wednesday, 6th.

From Taunton by railway (" The Flying Dutchman,") to Bristol : thence by another train to Gloucester, and then in Mr. Gambier Parry's carriage to his house at Highnam, about three miles off.

This house is quite a museum of mediæval 1876. decorative art, and of pictures by the early masters ; the shortness of our stay did not allow us to see half the objects of interest—still less to remember them. There are a few modern pictures:—in particular, a charming portrait of Mr. Parry's grandmother, by *Romney*, and a capital portrait group by *Zoffany*, of three naval officers, one of them the famous Kempenfelt.

A large family party at Highnam:—Mr. Gambier Parry's son* by his first marriage, with his very pretty and interesting young wife Lady Maude, (Lord Pembroke's sister), and their baby.

———

Thursday, 7th.

A very interesting day, though the weather was excessively bad.

In the morning first, to see Mr. Gambier Parry's own Church of Highnam, at a very short distance from his house. It was built entirely at his expense from the designs of M. Woodyer? It is a very graceful structure, the spire particularly beautiful; interior painted in fresco by Mr. Parry himself, and truly admirable. I was especially charmed with a series of figures — a sort of procession—of Apostles, Evangelists, holy women, and others mentioned in the New Testament ; these are painted with exquisite grace and refinement. We did not think the church too dark, though the glass is all coloured.

Then to Gloucester with Mrs. Parry and Arch-

* Hubert Parry, the composer of the oratorio of " Judith."

1876. deacon Lear, who showed us the Cathedral. A
noble building indeed ; remarkable, not only for its
grandeur and beauty, but for the many different
times in which its several parts were built, and con-
sequently the many various styles of architecture
exemplified in it. In this respect it is almost
the direct opposite of Salisbury. According to the
English Cyclopædia, the crypt, the chapels surrounding the choir, and the lower part of the nave,
were built between 1058 and 1089 ; the south aisle
and transept in 1310-1330 ; the cloisters in 1351-
1390 ; the Lady Chapel towards the close of the
15th century ; and the central tower somewhat later
still.

The nave, Norman,—simple, severe, and grand,
with plain massy pillars and little ornament. The
choir Perpendicular, very rich and splendid with
a very beautiful roof. East window magnificent,—
I think the highest and largest I have seen ;
indeed I rather believe it is the largest in England.
The great central square tower is very noble ; the
open work of its parapet and of its four-corner
turrets or pinnacles, remarkably beautiful.

The crypt, evidently the oldest part of the church,
very dark and gloomy ; pillars very thick, short and
massy ; arches very low.

Tomb with figure (cross-legged) of Robert Duke
of Normandy. Splendid monument of Edward II.,
erected about 1334 (Britton).

In the afternoon, though the weather was excessively wet and stormy, we went to see Mr. Parry's
Pinetum, situated about a mile or more from the

house, and quite separate from the park. It is 1876.
certainly the finest collection of the sort that I have
seen, and must be one of the finest in the kingdom.*
I mean of collections of conifers exclusively. The
situation is admirable :—on the upper part of a large
and bold hill, sloping steeply down to the vale of the
Severn ; and thus affording not only beautiful views,
but a variety of sloping ground and favourable
exposures for the trees. The soil too is good;
whereas the house and adjacent part of the park
are on the lias clay, this Pinetum is on a *heath* soil
overlying sand (of the Lower Oolite ?) The climate
too is certainly much milder than ours. Extent,
23 acres. The collection is I should think nearly
complete as to the known species and varieties of
Coniferæ, and the individual trees superb in growth.

Note of trees especially fine or rare :—

Picea bracteata ; quite new to me and very handsome ; the largest leaves of all the silver firs.

Picea Pinsapo ; extremely fine.

Picea "Apollinis;" Mr. Parry thinks this distinguishable from Cephalonica. I do not.

Picea nobilis—superb.

Picea lasiocarpa ; very fine.

Picea Orientalis ; quite new to me, and very fine.

Pinus insignis ; very fine and vigorous, quite un-
hurt by frost (with *us* it will not stand the winter at
all).

Pinus ponderosa ; superb.

Cedrus Atlantica.

* I have never seen Dropmore since 1849, so cannot judge fairly of it.

1876. Araucaria imbricata seems to thrive particularly well here; the trees of it are very numerous and thriving, some very large; indeed there are more of them than I much care to see.

Friday, 8th.

From Gloucester to Hereford, where change lines, and so to Mountain Ash, the station for Duffryn. Journey altogether about 5 hours. Between Ross and Hereford, cross the Wye repeatedly, as it winds extremely: it is here a pretty river, but not a great one. From Hereford by Abergavenny, Pontypool, a beautiful country:— bold but not savage mountains, rich wooded dells, green pastoral valleys, bright sparkling streams.

Abergavenny is in a particularly beautiful country, at the junction of three delightful valleys, bounded by fine mountains.

Pass over the Crumlin Viaduct, a very lofty and very remarkable bridge (210 feet high) spanning a deep and beautiful wooded valley, (that of the Ebbw or Ebwy river) down which there is a delightful view from this elevation.

From Ross to Hereford, from Hereford to Abergavenny, and thence nearly to Pontypool, the whole country is of Old Red Sandstone, making itself, in general, very apparent by the deep red colour, conspicuous everywhere in the ploughed fields, broken banks, lane sides, and such like situations. A little before Pontypool, we enter on the Coal formation. Sandstones of this formation strikingly different in colour from those of the Old

Red, even where not stained by Coal. The country 1876. very hilly and extremely pretty, but here and there disfigured by coal-pits, steam engines and refuse heaps, and more often by hideous rows of cottages.

The Aberdares' carriage met us at Mountain Ash Station, and there, also, we met Catty and Harry Napier, who had come by the same train from Hereford, but whom we had not yet seen.

Saturday, 9th, to Tuesday, 12th.

At Duffryn (or Dyffryn, as it is spelt in the Ordnance Map).

The weather was very bad while we were here, and I was not well, which rather interfered with my enjoyment, yet we spent these four days very pleasantly. Aberdare himself was unfortunately absent, as well as his two eldest sons, Harry and William, but Norah is always delightful: indeed, I know very few, if any, women more to be admired, and her sister Caroline is hardly less charming. The two eldest girls, Caroline (Lina) and Sarah, are very pleasing, and the last day and half, Mr. Clark (of Dowlais), was a guest in the house. Mr. Clark, of Dowlais, spent some days at Barton in July, 1869—see my journal of that date. He is a remarkably well-informed and agreeable man :— his *speciality* is archæology, and in particular, all that relates to ancient castles and fortifications. The Aberdares have a very high opinion of him. There were plenty of excellent books in the house, and pretty views to be seen whenever we were

1876. able to go out. The house of Duffryn has been so very much enlarged, altered and improved, since we were there in '64, that it appears quite new. It is now indeed a very handsome and very good house, and the grounds also have been extended and much adorned.

Duffryn is situated in a long narrow and deep valley—that of the Cynon (pronounced *Cunnon*), a tributary of the Taff—between two ranges of tolerably high and very steep mountains, composed of Carboniferous rocks. The bad weather, and my health, prevented me from ascending any of the hills this time: but in September, '64, I ascended one of those nearest to Duffryn; I found the upper part of it clothed with short grass and a profusion of dwarfish Bilberry bushes: very little heath and no abundance of Moss. The table-land seemed to be extensive.

The general direction of the valley is from N.W. to S.E. or thereabouts. The town of Aberdare is near its head. The general character of the hills or mountains is rather uniform: they have steep sides, generally well wooded and broad flattish tops, showing long straight sky-lines, their upper parts have generally a smoothish appearance, not showing much rock or stone.

On the 9th, in the afternoon, we went out in the carriage with Norah and Catty, a long drive through some very pretty wooded valleys. It appears as if the scenery of this district must have been really very pretty before it was interfered with by coal-mines and steam-engines, and, worst of all, by the

peculiar hideousness of the cottages. In our drive, 1876.
Norah pointed out to me a rather conspicuous crag,
named Daran-y-cigfron (I copy the spelling from
the Ordnance map), overlooking the valley of the
Taff.—But in general, abrupt and precipitous rocks
of large size seem to be rare among these hills.

Mountain Ash, which is in the valley, and extends
to within a few hundred yards of the grounds of
Duffryn, is a town of about 9,000 inhabitants, and
one of the very ugliest little towns I have ever seen.
Nothing but coal-pits, iron works, huge black
machines, cottages of peculiar hideousness, and
everything—animate and inanimate—begrimed with
coal dust and black as chimney-sweepers. I have
heard Lord Aberdare say, that he very well
remembers the site of Mountain Ash, when there
was only one solitary house on it,—a small inn with
the sign of the *Mountain Ash*. Now it is a town of
9,000 inhabitants.

The coal of these valleys is anthracitic—not pure
anthracite however, but with a very small proportion of bitumen. Hence, though the ground,
the houses and the people are begrimed with coal
dust, there is no such perpetual mantle of black
smoke hanging over the towns and villages, as one
sees in the colliery districts of Staffordshire and
Lancashire.

About three weeks before we were at Duffryn,
during a great thunderstorm, a formidable torrent
came down with extreme suddenness from the
mountain, carried away great part of the road
between Duffryn and Mountain Ash, blocked up the

1876. approach to the town with huge stones, and in
short, did enormous damage, though no life was
lost. Norah pointed out to me many traces of its
havoc still remaining.

Wednesday, 13th.

We left our kind friends at Duffryn, went back as
far as Hereford by the same line by which we had
come, and from Hereford to Great Malvern.
Beautiful country between Hereford and the
Malvern Hills. A great tunnel near Ledbury
(through some of the Silurian Hills, I suppose), and
another through the Malvern Hills themselves,
coming out between the Wells and Great Malvern.
The *Foley Arms* is in an excellent situation, but the
evening was so foggy that we could see little of the
view.

Thursday, 14th.

From Malvern by the Great Western Railway, by
Worcester, Evesham, Oxford and Reading (without
any change of train), to Paddington. We arrived
safe at 48, Eaton Place about 5.—Thank God.

The MacMurdos dined with us. Very pleasant.

Friday, 15th.

48, Eaton Place.

Dear Arthur went back to school. Fanny going
with him to the railway.

Minnie and Sarah dined with us.

George came to luncheon.

The MacMurdos (including Mimi), Minnie and her brother John dined with us.

———

Monday, 18th.

We went with the MacMurdos to the National Gallery, and through all the new rooms, which are very beautiful, and in which the pictures are shown to great advantage. Afterwards to Minnie's : saw her and Sarah and dear little Charlie—a lovely child.

———

Tuesday, 19th.

Lady Head came to luncheon. Cissy came to stay, and George dined with us.

— — —

Wednesday, 20th.

The MacMurdos came to luncheon. Mrs. Douglas Galton dined with us. A visit from Sir Francis Doyle.

— — —

Thursday, 21st.

Dear Cissy left us to go back to Folkestone.

I went to say good-bye to Minnie (who was not well), and Sarah came to luncheon. We went down to Barton by the 4.20 train from Liverpool Street, arrived safe and found all well at home.— Thank God.

1876.   Friday, 22nd.

Miss Spencer, Mrs. Hamilton Grey and Lady
Frederick Beauclerk came to luncheon and to see
our pictures.

_____

Tuesday, 26th.

A long visit from Mrs. Horton, Lord and Lady
Tollemache and Lady Florence Barnardiston.
Clement arrived.

=======

LETTER.

Barton,
September 26th, 76.

My Dear Katharine,

A thousand thanks for your letter fro
Gairloch Hotel, which I received this morning, and
have read with great delight. I am quite charmed
with your account of your expeditions and adventures in Skye, an island which I have always
thought particularly interesting—associated in my
mind with *Eriocaulon* and with Macculloch's min-
eralogical descriptions, as well as with "The Lord
of the Isles," and Johnson's Tour. We have lately
also been hearing a great deal about it from another
quarter for MacMurdo and his daughter Mimi, and
Lord Aberdare were staying some time at Dunwegan
Castle; and were delighted with what they saw.
How fortunate you were in weather!—a very rare
felicity, I should think, in that island. We were in
a very different predicament—persecuted all through
our tour by bad weather, which especially interfered

with our enjoyment at Lynton. I am delighted to hear that Rosamond is so strong; how she must have enjoyed her rambles in that beautiful country, and among those wild scenes of Nature. I am extremely glad you found the Eriocaulon, which I never had the luck to see living : but it seems to dry very well. All the other rarities which you mention, I have found in other places. Did you find the Subularia, or the Isoetus ?

I found nothing new to me in our tour, not even a moss or a lichen; but it is always pleasant to notice the characteristic plants of each district, when they are different from what one is used to at home ; such as the Origanum and the Viburnum Lantana, on the chalk of Kent, and on the oolite about Sherborne, and the Golden-rod, the Foxglove and the great Wood Rush in the woods of North Devon.

I think Fanny will have told you all about our tour which (in spite of the wretched weather), was very interesting and very pleasant. Arthur contributed a good deal to the pleasure of it by his unfailing good spirits and good humour, and his quickness of observation ; and he made friends wherever he went. The parts of our tour which were most new to me were Sherborne, Gloucester, and the country between Hereford and Mountain Ash (this last place, which you will not easily find in the map, though it is a town of 9,000 inhabitants, is near Aberdare, and is the station for Duffryn).

I had great pleasure in seeing again Wells, Salisbury, Stonehenge, and especially Lynton. This

1876. last I have seen now four times, and I was as much charmed with it this last time as ever. It was a great pleasure to see so much of the Sanfords, and to see Sarah in her home ; I wish I could add that I thought her looking well. Mr. Sanford was extremely agreeable, and the young people very pleasant. Woody Bay is a most beautiful spot, but the way to it from Lynton is (as Fanny will have told you) rather perilous looking.

Now we are quiet at home again, and not sorry to be quiet. We have already had guests ; Mr. and Mrs. John Martineau, great friends of the Kingsleys, and of Mr. Sanford, came to us on the 23rd, and have just left us. They are very agreeable and very likeable people indeed. Mrs. Martineau was one of the Miss Adeanes, a cousin of the Blakes ; Mr. Martineau was a pupil of Mr. Kingsley's at Eversley, about the time when he was writing Alton Locke.

I hope you will come and stay with us some time this autumn. I do not think you have been with us since the plants which Dr. Hooker sent us from Kew were placed in our new hot-house ; these are now growing beautifully, and so are most of those which we bought at Veitch's in June.

With much love to Rosamond and Arthur, and Fanny's best love to all of you,

Believe me ever,

Your affectionate brother,

CHARLES J. F. BUNBURY.

JOURNAL.

We (including Clement) dined with the Hortons : 1876.
met the Barnardistons, Mrs. James and her son,
Miss Tollemache, Herbert Praed.

———

Thursday, 28th.

Wet weather—one shower very heavy indeed.
My Barton Rent Audit, and luncheon with the
tenants and Mr. Percy Smith afterwards.

———

Friday, 29th.

Colonel Ives and Mr. Thornhill came to luncheon.

———

Sunday, October 1st.

We went to Morning Church and received the
Communion.

———

Thursday, 5th.

Fanny went alone to dine with the Iveses at
Rushbrook, as I had a cold.

———

Sunday, 8th.

Immediately after our return from London (from
the 23rd to the 26th), we had the very agreeable
company of Mr. and Mrs. John Martineau, great
friends of the Kingsleys, and of Mr. Sanford. Mr.

1876. Martineau is a particularly well-informed and agreeable man.

The remarkable meteor of September 24th was seen by several people here, and excited great astonishment. We unluckily were not out of doors, nor looking out at the time, and missed it.

Since our return home we have remained, in general, very quiet, and seen but few people.

Letters have come in these last few days from Susan, from Munich, giving us bad news of poor George Pertz, who has had a paralytic stroke, and is in a very critical state. He is above 80 years old. I am very sorry for Leonora and for their daughters.

The aspect of political affairs is extremely gloomy, uncertain and threatening. There appears great danger of our being forced, even in spite of ourselves, into a dangerous and ruinous war, for an odious cause. A war for the purpose of upholding the Turks and enabling them to maintain their tyranny, would, in my view, be a national crime and misfortune. But neither can I at all approve Gladstone's scheme of a crusade against the Turks. I have no confidence at all, either in Mr. Gladstone or Lord Beaconsfield ;—some in Lord Derby, Lord Caernarvon, and perhaps Lord Hartington.

LETTER.

Barton.
October 9th, 1876.

My dear Katharine,

Very many thanks to you for sending me that extract from Mary's letter about the *great event* of May 30th, '44 ;* it has interested me exceedingly, 1876. and I shall keep it as a most valuable record. It is, as you say, an old story now, but it is one which I neither am likely to forget nor have any wish to forget, and it is very pleasant to have its particulars recalled by so lively and well-written an account of it as dear Mary's. How well I remember that breakfast! How young and beautiful Mary herself looked then! I do indeed feel strongly what a blessing it is that we can look back on these thirty-two years with thankfulness for the happiness we have enjoyed in each other's society, and for the unchanged and unabated continuance of it to the present moment.

I have been much grieved by the news of George Pertz's illness, and feel great sorrow for Leonora and the poor girls. I should fear that at his great age a real recovery is not probable, and if that does not take place, I hope our poor old friend will not linger long in a disabled and helpless state.

I hope you received my letter of the 26th September, directed to Sheilhill, in answer to your delightful one from Gairloch.

I was much interested by your accounts of what you had seen and done in Skye, and wished I had been with you both there and at Clova. I am looking forward with great pleasure to seeing you and dear Rosamond here, and only wish you would stay longer with us.

Ever your affectionate brother.

CHARLES J. F. BUNBURY.

---

* An account of his wedding. *See* June, 1844.—F. J. B.

1876.   *October 13th.*—I had written this before we received the news of George Pertz's departure from
this world.   I had hardly expected that the end
would come so quickly, but no doubt it is best ; if
he was not to recover, one could not, even for his
own sake, wish him to linger on in a half-living
state.   At least, I am sure that I, in such a case,
should much rather wish to die.   I feel very much
for poor Leonora and the girls, to whom it must be
a terrible grief; though indeed Leonora must have
been in some degree prepared by his age and
his infirm health of late years ; still I dare say she
did not expect it just then.   He was a very good
kind man, and must be dreadfully missed in his own
family ; and he had been a most useful and laborious man of letters.

I hope this will not prevent you from coming
to us.

JOURNAL.

Tuesday, 10th.

Charlotte and Octavia Legge, the Montgomeries
and Mr. Holmes arrived.

The Horton party (4), Agnes and Arthur Wilson
and Miss Maitland dined with us.

We received the news of the death of our brother-
in-law George Pertz.   The end has come rather
more quickly than I expected ; it was only on the
5th that we heard of his paralytic attack.   He died
at Munich, surrounded by his wife and children and

sisters-in-law. He was a very kind and good man, 1876. and had been one of the most industrious and useful literary men of Germany.

—————

Thursday, 12th.

A beautiful morning, very mild. I out first time for several days—went round the garden and arboretum.

Miss Gibson and her brother Gery, Mr. and Mrs. Percy Smith dined with us.

Weather for a week past or more, remarkably warm as well as excessively damp, so that, as I hear, the grass is growing faster than in the spring.

On the 7th (Dr. Macnab told me), at noon, at Bury, the thermometer was at 71 degrees in the shade.

—————

Friday, 13th.

Mrs. Ellice and Helen arrived late.

—————

Saturday, 14th.

Mr. and Mrs. Montgomerie and Mr. Holmes went away.

A walk in the Park with Fanny.

The Louis Mallets arrived.

—————

Monday, 16th.

The two Lady Legges went away. Lady Head and Amabel arrived.

—————

Tuesday, 17th.

Showed the new plant-houses to Lady Head.

Received a note from Charles Bell (Sir John's nephew), whom I knew very well and saw often at the Cape, but have never seen, and scarcely heard of since those days,—that is for 38 years. It seems he has now come to live with his uncle (who is very old, 95 or more), and to take care of him. I am very glad to renew acquaintance with one who is associated in my memory with those delightful days at the Cape.

Wednesday, 18th.

Wrote a long letter to cousin Emily (Ambleside).

The Milner Gibsons, Mr. James Bevan, Headley Bevan and the Percy Smiths dined with us.

———

Thursday, 19th.

Most beautiful weather—as hot as summer.

Poor Calthorpe called away by the illness of his child.

A pleasant walk with Helen Ellice and Miss Grant. Helen and Mrs. Ellice went away in afternoon.

———

Friday, 20th.

Herbert arrived.

———

Monday, 23rd.

Katharine and Rosamond arrived.

Resumed the arrangement of Charles Lyell's 1876. letters.

Very pleasant company here the last fortnight :— Charlotte and Octavia Legge, the Montgomeries and Mr. Holmes, Louis and Fanny Mallet, Lady Head and her daughter Amabel, Mrs. Ellice and Helen and dear Kate and Charles Hoare.

We have also entertained a good many of the neighbours. One afternoon we had an unexpected visit from the Duke of Grafton and Lord Hertford, both of whom were extremely agreeable. Lord Hertford seemed much struck with our pictures. The Mallets, Heads, Ellices were particularly agreeable. Louis Mallet is looking well, is in very good spirits and seems to have quite recovered from his Indian illness. He is a man for whose character and abilities I have a very great respect : and he is not only good and wise, but his con- versation among his friends is lively, various and entertaining. He has much humour, especially in telling stories. His wife also is a favourite of mine.

The Montgomeries are very pleasant, and the Lady Legges, especially our old-young favourite Octavia.

———

Tuesday, 24th.

Susan and Mimi MacMurdo arrived ; and in the evening, Lina, Sarah, and Harry Bruce, Mr. Egerton Hubbard, and John Hervey. Katharine gave me a specimen of Eriocaulon.

Katharine has shown me a letter she has received

1876. from Lady Smith, from Lowestoft; really a very extraordinary letter to be written by a woman in her 104th year. Her sight is somewhat failing, and her handwriting therefore rather illegible, but her mind is still clear and vigorous, and she takes an intelligent interest in what is going on.

Katharine is lately come from Scotland, where she (and Rosamond and Arthur with her), spent a considerable time very much to their satisfaction. After attending the British Association meeting at Glasgow, they joined with Dr. Hooker and his bride and father-in-law, in a visit to the Isle of Skye, with which she was very much delighted. The Hookers and the younger members of the party explored the wilder mountains and lochs, while Katharine and Sophy Lyell enjoyed the more accessible beauties. Katharine speaks with enthusiasm of the beauty and wildness and singularity of the scenery, and especially of the bold outlines of the mountains. Katharine had the satisfaction of gathering the rare Eriocaulon septangulare in the Isle of Skye, and has given me specimens of it. She and her party had the good fortune of enjoying almost uninterrupted fine weather, without either wind or rain, during their stay in Skye; very uncommon good fortune indeed, I should think, in that island so famous for storms.

Wednesday, 25th.

Went through the plant houses and arboretum with Katharine and Susan MacMurdo.

Mr. and Mrs. Montgomerie and my nephew Harry

arrived; Agnes and another Miss Wilson, and Mrs. 1876.
Victor Paley dined with us.

Showed the arboretum, &c. to Mr. Hubbard.

Montagu MacMurdo arrived.

Most of our party went to the Bury Ball, but
poor Mrs. Montgomerie was ill, and not able to go.

The delightful news of dear Sarah Seymour's
safe delivery, and the birth of her little daughter—
"mother and child doing well." It is nearly two
years since the birth of her first child Charlie.

The Charles Hoares, Katharine and Rosamond,
Mr. Hubbard, two Mr. Hoares, John Hervey, Harry
and Clement went away.

Received a very interesting letter from Edward,
finished at Corfu, but principally written at Cattaro.
He had made an expedition from thence, in company with the Italian consul, to Cettinje, the capital
of Montenegro, and found (as he had been told),
that there was no danger, at least from the people,
though much of roughing and discomfort. A battle
was actually going on within 20 miles of Cettinje the
very day that he arrived there, yet there seemed to
be a total absence of marauding and of irregular
violence. The letter gives a striking description of

1876. the country, and interesting remarks on the people with whom Edward seems to have been very much pleased. They appear in many respects to be in a similar stage of civilization with the Scotch Highlanders of *Waverley's* time, but some of the war customs of the Montenegrins are more ferocious than those of the Gael ever were. They collect the *noses* (as the Red Indians do the scalps) of their slain enemies, as trophies of victory.

The newspapers give the surprising news of the return (unexpectedly early) of the Arctic Expedition. They seem to have satisfied themselves that the Pole is entirely inaccessible, but that there is no land near it. I suppose it is surrounded, like the South Pole, by an impenetrable barrier of solid ice. A few particulars, not many, are given in the telegrams, but as the expedition has reached Ireland, we shall soon know more. They reached a higher latitude than any (at any rate civilized), men had ever arrived at, namely, 83° 20', and this "after a terrible struggle," they were then 450 miles from the Pole. They were also exposed to the " severest weather ever registered, the temperature being 59 degrees below zero for a fortnight, and the extreme lowest temperature on any day being 104 degrees below freezing point " (of Fahrenheit I presume). The ice was so rugged that the sledgers could only advance *one mile* a day. Thus it appears clear that to reach the Pole was (as Caplain Nares says), quite impracticable : and I hope that we shall hear no more of Polar expeditions.

There are many further particulars of the Arctic expedition, but not giving any ideas materially different from what we received at first. It is clear that the hardships, sufferings and dangers, especially of the sledging parties, were excessive, and it seems really marvellous that any of those brave men returned alive. All honour is due to them, but I would rather not hear of any others following their example.

It is said that many specimens of natural history have been brought home, and a great many valuable observations made by the expedition.

The Montgomeries and Sir John Shelley went away.

The Barnardistons, Admiral Spencer, Caroline Newton and her niece Ada arrived.

Out all the morning with Fanny and Scott marking trees.

———

Tuesday, 31st.

Showed some of our minerals to Lina and Sarah Bruce.

Mrs. Horton and Lady Winchelsea came to luncheon and to see our pictures.

Lady Parker and two daughters arrived.

Colonel Nason dined with us.

———

November 6th.

Mr. Harry Jones, our neighbour at Barton Mere, has made us a splendid present of a plant of Agave

1876. Americana, just beginning to push up its flower stem. He has no room for such a plant in flower, and therefore has transferred it to us, as in our new hot-house it will have a chance. I shall be much interested in watching its progress. That new hot-house has very lately brought us two other presents, from Mr. Montgomerie, a fine tall plant of Ficus elastica, and a beautiful tree fern, whether Cyathea or Alsophila I cannot tell without the fructification.

The numerous little seedling plants (many of them rare and curious), which Dr. Hooker sent us from Kew, in the early part of the year, are thriving.

_____

November 8th.

We are now quite alone, for a wonder ; Lina and Sarah Bruce, who stayed the last of our party (stayed very kindly to help Fanny in writing), left us on the 6th.

Alice Praed is engaged to be married to Mr. Holden, the clergyman at Lackford, a very intelligent and gentleman-like man, who has several times been here. She is a very nice girl, and I am very glad she has the prospect of a happy marriage.

_____

LETTER.

Barton, Bury St. Edmund's,
November 9th, '76.

My Dear Katharine,
        I am very glad you enjoyed such a

pleasant time with the Hookers, and in those ever 1876. delightful gardens. The *Lapageria* I know very well, and have seen in very great beauty in Mr. Berners's conservatory at Woolverstone, near Ipswich, and also at Veitch's, but it is difficult to manage, and we have not yet been successful with it. At Woolverstone they have the white variety as well as the red, and there, also, I saw it bearing *fruit*—unripe however, and something like a very young smooth cucumber. The Rhodochiton I remember seeing long ago, but we have it not.

Schimper's work on Mosses, newly published, is only a 2nd edition of his Synopsis, which I bought 15 years ago, and have studied and used a good deal, and like very much, except that he multiplies both genera and species most inordinately. All the introductory part, all relating to the localities and geographical range, &c., of Mosses, is excellent.

Captain Barnardiston has given us a plant of a rare Orchid from Panama, the *Flor del Espiritu Santo*, a Peristeria. And Mrs. Horton has given us an Adiantum, but I suspect it will prove only a variety of *trapeziforme;* at present it is not in fructification.

Ever your loving brother,
CHARLES J. F. BUNBURY.

JOURNAL.

November 11th.

We received the news of the death of our very old

1876. and dear friend Lady Bell. She was in her 90th
year, and for some weeks past has been rapidly
failing, so that there could be no doubt as to what
was coming : but her mind continued clear to the
last, and her feelings as warm, affectionate, and
true as ever.   Hers was a beautiful life and a
beautiful death.

My acquaintance with Lady Bell began in 1831,
through the Napiers, but she was much earlier an
intimate friend of Fanny's family.   She had been a
friend of Sir George Napier's wife, and thence was
particularly attached to Cissy and Sarah and their
brothers.   Her soft caressing manner was the true
indication of a wide and most tender, most loving
heart.   It would be difficult, I should think, for any
human being to be more perfect in all moral
qualities.

_____

November 21st

Another week of company and of pleasant dissipation, from the 13th to this day. A special assemblage
of delightful people ; Minnie (our own dear Minnie),
and Minnie Powys, W. Napier, Cissy Goodlake and
her husband, Sir Edward and Lady Greathed,
Isabel Hervey, Mr. Yeatman, John Herbert, Lord
John Hervey.

_____

November 22nd.

I see in _The Times_ that my old friend Gen. Sir John
Bell died the day before yesterday in the 95th year
of his age.   I knew him well at the Cape of Good

Hope, where he was Colonial Secretary when Sir 1876.
George Napier was Governor ; and as, besides this
official relation, they were old friends and fellow
soldiers, Sir John was continually at Government
House, and I saw him almost every day. He was a
fine old soldier, a distinguished member of the
famous Light Division, and actively engaged
throughout the Peninsular campaigns. A good
specimen of a Scotchman ; shrewd, cautious, observant and thoughtful ; with a great deal of dry
humour ; and with great kindness of heart. For
many years past he has lived chiefly in London,
where we have visited him every year, and he has
always received us in a very friendly way. He has
however latterly been so deaf, that conversation with
him was difficult. It was in the 52nd regiment,
I think, that Sir John Bell served in Spain and
Portugal ; and I have heard that he was especially
remarkable for his talent for military drawing, for
reconnoitring and sketching positions, and the like.

He married Lady Catherine Harris. I remember
her very well at the Cape ; a stately lady, with
an old-fashioned dignity of manner, and something
of old-fashioned rigidity in her opinions ; a Tory of
the straightest school, and much inclined to believe
in the progressive degeneracy of mankind. Yet she
was exceedingly kind (as was her husband also),
to Sir George's daughters, and indeed really acted
the part of a mother towards them, until their
father's second marriage.

We returned yesterday from a two days visit to the Angersteins, at Weeting. The house is large and good and comfortable ; interesting only on account of the pictures, which are very numerous, and several of them very good. There are *Sir Joshua Reynolds's* famous "Garrick between Tragedy and Comedy;" his charming "Piping Boy" (a naked nymph reclining, listening to a piping swain) ; several excellent family portraits by *Reynolds*, *Romney*, and *Sir Thomas Lawrence* (indeed some of *Lawrence's* pictures at Weeting appear to me the best of his works that I have seen ; some admirable works of *Turner*, in his best time.

LETTER.

Barton,
November 27th, 76

My dear Cissy,

I was very happy to hear of your prosperous journey to Cannes, and your safe arrival there, that dear Emmie was not the worse for the journey, and that the place seemed to agree with her. I hope and trust that it will do her a great deal of good. That your stay in the South will be both pleasant and profitable to the health of both of you, and that when you return to England, Emmie will be in thoroughly good health.

You may perhaps before this have seen in the newspapers that our old friend Sir John Bell is dead

—in his 95th year, the papers say. One cannot be surprised at it, at that age, and I do not know that one ought to regret the death of so old a man, for whom life could have little enjoyment; who could in fact have little to live for. But yet I do feel some regret at the breaking of one of the few remaining links with those delightful Cape days, which I constantly recall with so much pleasure, and I think you will feel the same, perhaps more than I do. John Bell was a fine old soldier, a hearty, steady friend and a good man, humorous and pleasant.

On the 17th of last month, I was surprised by receiving a letter from *Charles* Bell* (whom I had neither seen nor heard of since we left the Cape), recalling himself to my memory and telling me that he had come to live with his uncle and take care of him. I wrote immediately a friendly answer; and on seeing Sir John's death in *The Times*, I wrote to Charles Bell to condole with him ; but I have not heard again. I daresay he is very busy. There have been many deaths lately among our old friends or connections; first, our brother-in-law George Pertz, then dear Lady Bell, then Fanny's aunt, Mrs. Byrne, then Sir John Bell, and almost at the same time, old Mr. Prevost of Geneva.

We spent two days of last week in a visit to the Angersteins at Weeting; they are friendly people and have a very good house, containing a multitude of excellent pictures.

The company were a *shooting* party, not congenial

* Nephew of Sir John Bell.

to me ; the pleasantest, Lord and Lady Henley,— the latter a friend of the Aberdare Bruces, and particularly of Rachel Vernon Harcourt.

I am reading, and have nearly finished, the Memoirs of Mr. Ticknor, an American, who wrote the Life of Prescott ; he was a great friend of the Lyells, and seems to have known everybody that was worth knowing, in every part of Europe ; his memoirs are very entertaining.

I am every day expecting the Life and Letters of dear Charles Kingsley, of which Mrs. Kingsley has promised me one of the first copies.

My best love to dearest Emmie, and Fanny's to both of you.

Ever your very loving brother,

CHARLES J. F. BUNBURY.

JOURNAL.

November 30th.

I see in *The Times* the appointment of Sir Bartle Frere to be Governor of the Cape of Good Hope.

December 6th.

I received from my dear friend Mrs. Kingsley (sent by her publisher at her desire), the two volumes of her husband's "Life and Letters," edited by her. Read the first three chapters. The book begins delightfully, and is sure to be exceedingly interesting, to me especially. Her dedication (to *his* memory), and preface are beautiful.

Finished, on the 21st, the 1st vol. of Kingsley's "Letters and Life;" wrote to Rose to tell her my impressions of it, which I will also put down here. All that Mrs. Kingsley has done—all her share of the book—I think admirable ; indeed, I am quite struck with the merit—the unaffected beauty—of what she has written. Of Kingsley's own letters, in this first volume, a good many of those on theological subjects, are to me very obscure : some quite incomprehensible, full of high-flown mysticism and Platonic metaphysics, which convey no ideas to me. I do not doubt at all that he had ideas in his own mind corresponding to them. Probably the deficiency is in *me*. But wherever he speaks of what I call *religion* proper, and not metaphysics, I recognize that beautiful and elevating tone which impressed me so much, not only in his sermons, but in all I have heard him say on those subjects.

Altogether this volume is exceedingly interesting, and is far from disappointing me : and so far from lowering him, even raises him in my estimation, highly as I admired him before. Nothing can be more beautiful than the unselfishness of his character, his constant self-sacrifice and devotion to the good of others. His intense trust in God, and unwavering belief in His goodness and His support, are most noble, and his letters are full of valuable lessons in this way. The letters printed here show how many people, of most various positions and pursuits, derived comfort

1876. and spiritual aid from him while he lived, and I am sure this book will be valuable to very many.

On the question of Future Punishment, and the doctrines of Calvinism as connected with it, Kingsley expresses himself with the utmost plainness and vigour and unsparing energy (see particularly pp. 470, 471). I suppose he would be considered as a heretic on this ground; for my part, I admire him most heartily for it, and would gladly subscribe to every word.

During a great part of the time comprised in this volume, especially from 1847 to 1856, Kingsley's thoughts and energies were very much absorbed by the questions relating to the condition of the lower classes. These were the days of his "Christian Socialism," of Yeast and Alton Locke.

All the correspondence on this subject is very interesting. I can perceive in these letters, what Mr. John Martineau told me, when he was with us last September—from his own personal knowledge —that for some years Kingsley hardly thought of any other subjects. My acquaintance with him began some years later, when he had calmed down, and from various causes had lost something of the vehemence and asperity of those early works, and allowed himself to give more free play to that variety and "many-sidedness" of mind, which was one of the things that made him so delightful. One excellent specimen of this quality, indeed, we have near the end of the 1st vol., in the charming account of his expedition with Mr. Hughes and Mr. Taylor to Snowdon;

p. 493 to 496 of this volume. In this rapid rattling 1876.
sketch we can see the origin both of a chapter of
Two Years Ago, and of the first three pages of
Chalk Stream Studies, in the Prose Idylls.

This is a fine specimen of the union of rollicking
joyousness with accurate observation.

Of the letters to Mrs. Kingsley from friends,
contributing their views of her husband's character,
there are two in this volume which are particularly
good:—from Mr. C. K. Paul (p. 224-230), and
from Mr. John Martineau (p. 297-308). This latter
especially seems to me an admirable picture of the
man.

LETTER.

Barton,
December 26th, 1876.

My dear Leonora,

I am very sorry that (I do not know how),
I neglected to write to you when you suffered so
great a loss in the death of your good husband. I
am sure I did feel for you, and I was well aware
that you were deeply attached to him, and that
his death made a very great change to you. Indeed
it must have been widely felt, for he was a very kind
and good man, and his great and useful and valuable
works on German history will always preserve his
name. But it must be a comfort to reflect that
at that age, if his life had been much prolonged, it
must probably have been accompanied with evident
decay and the failure of vital power and of much

1876. enjoyment; and I am sure you have the comfort of feeling that you made the last twenty-two years of his life very happy and smoothed his passage to the tomb. I am glad you are now at Florence with Susan and Joanna, it must be cheerful both for you and them, and extremely interesting to Annie and Dora. You all seem, by the letters Fanny reads to me, to be extremely industrious. I hope you will come to us this summer, and I think your two dear girls would find some things here which might interest them. You will see something which was not here when you were last with us. I think you would like to see our new plant houses.

With much love to Annie and Dora, Susan and Joanna, and all possible good wishes to all of you for the new year,

Believe me ever,

Your affectionate brother,

CHARLES J. F. BUNBURY.

JOURNAL.

December 28th.

The year which is now near its close has been much less fatal—at least in our own circle, than '75. The relations and intimate friends we have lost have not been numerous, and with one or two exceptions, we have been prepared for the loss, either by their great age or their infirm and helpless condition.

I may mention here two or three eminent men

with whom I was acquainted, but not intimate, who have died in the course of the year:— Adolphe Brongniart, the Linnæus of fossil botany, Paulett Scrope, — perhaps I may add Sir John Coleridge.

I have again to acknowledge with deep gratitude, God's constant goodness to me, in allowing me still to enjoy so many blessings; above all, the companionship of my dear wife, and her and my own good health, and the continued attachment of so many valuable friends. Our wordly circumstances are comfortable.

The aspect of public affairs is very unsettled, dark and threatening. Happily, this applies only to foreign politics, for at home, there is general tranquillity and contentment. But the "Eastern Question" is a terribly threatening complication: and though the Conference has met, and England is well represented by Lord Salisbury, who will probably do all that can be done to get us out of the difficulty, there still seems to be an awful probability of a great war. It is a terrible prospect, even if the war were to be only between Russia and Turkey; even this would involve a shocking amount of slaughter and misery: but I fear it would be found very difficult to "localize" the war, and it would be frightful if we were to be dragged into it. To be involved in a horrible waste of life, a waste of wealth, and all the resources of civilization, and exposed to unknown dangers for the sake of upholding Turkey. What a misery! May God preserve us from such a war.

1876. May He "grant unto all nations unity, peace and concord." I do not think there is anything else so much to be prayed for.

——————

This year has been peculiarly fertile in interesting and valuable biographical Memoirs.

——————

BOOKS I HAVE READ IN 1876.

——

Life of Lord Palmerston, by Evelyn Ashley. 2 vols.

Life of Macaulay. 2 vols.

Life of Norman Macleod. 2 vols.

Lord Albemarle's Reminiscences. 2 vols.

Life of Lord Althorpe. 1 vol.

Memoir of Caroline Herschel. 1 vol.

Life of George Ticknor. 2 vols.

Life of Charles Kingsley. 2 vols. (not yet quite finished).

Kinglake's Ivasion of the Crimea. Vol. 5.

Freeman. Historical and Architectural Sketches. 1 vol.

Wallace. Geographical Distribution of Animals. 2 vols. (A work of immense research and great value, but of course suited for consultation rather than for reading).

Moggridge. Flora of Mentone. 1 vol. (This also not read through, but examined and consulted.)

Moggridge. On Harvesting Ants and Trap-door 1876.
Spiders. 1 vol.
Lubbock. On British Flowers fertilized by Insects.
1 vol.

EMINENT PERSONS

Who have died in this year, with whom I was not acquainted.

Ehrenberg.
M. Mohl (a great Orientalist).
Miss Martineau.
George Sand (Mme. Duddenant).
John Forster (author of "Life of Sir J. Eliot,"
"Statesmen of the Commonwealth," &c.)
E. W. Lane ("Modern Egyptians.")
Colonel Chesney.
Sir George Bowles.
Sir Baldwin Walker.
Sir John Kaye.
Lord Lyttleton.
Cardinal Antonelli.
Marshal Saldanha.
Lord Tweeddale.

1877.

——◦◦——

JOURNAL.

1877.   Dear Minnie arrived, also Herbert.
The servants' supper party and dance.

———

A pleasant stroll in the gardens with Fanny and Minnie.

Arranged some more of Rose's Californian plants.

———

Meeting (in this house), of the Trustees of Howardley's Charity (Mr. P. Smith, Mr. King, Mr. Phillips, Scott and I),—satisfactory.

The good news of the birth of another child (a girl), of Leonard and Mary Lyell, and that mother and child are doing well.

The season is very extraordinary: constant mild weather through the whole of December and as far as January has yet gone, extremely few frosty days, nothing like severe frost, and no snow at all. Rains more continual than I almost ever remember. Here, we have had no remarkable violence of wind, but on the north-east coast, Northumberland and Durham, and also along the south coast, the gales seem to have been tremendous. The newspapers

are full of distressing accounts of the mischief done 1877.
by storms on the coast, and by floods in the low-
lying inland districts.

_____

Thursday, January 11th.

Miss Bucke, the Robesons, and Alice Praed
arrived.

Arthur's "entertainment"—a very numerous party
of young people — great gaiety and merriment,
dancing and games, &c.

_____

Friday, January 12th.

Examined some Ferns from Java.

_____

January 21st.

Much company and gaiety during the past week;
indeed very unusual gaieties, for I do not think that
this house has ever before, within my memory, witnessed a _fancy dress_ dance. Our party assembled on
Monday the 15th; when the Palmer Morewoods,
the Barnardistons, the Angersteins (with one
daughter and one son), Mrs. and Miss Byng, Isabel
Hervey and her friend Miss Herbert, William
Hoare, Lord Ernest Seymour, Mr. Anstruther,
John Hervey arrived; making up, with dear Minnie,
(who has been with us since the 4th), Herbert,
Harry, Arthur, and our two selves, twenty-two in
all. The 16th there was the fancy ball at Bury,
and those of our guests who were going to it (except
the _chaperones_ and John Hervey, who is a clergyman)

1877. appeared at dinner in their fancy dresses. The ball seems to have been a great success. Mrs. Angerstein, Arthur and I were the only ones who did not go to it. The 17th was Fanny's fancy dress party and dance, in which the dresses of the ball reappeared ; the number was about 80, and it was a very pretty and amusing sight : a great many pretty women, brilliant dresses, and much animation. Patience Morewood looked lovely in a splendid Syrian dress which she had brought from the East last year ; Isabel Hervey showed to great advantage in a very rich old-fashioned Court dress, which, she said, had been made for her mother. Miss Byng appeared as " Penelope, Boothby " after Sir J. Reynolds's picture : Lady Florence as "Mary Queen of Scots," in a superb dress : Barnardiston as " Darnley," not a bit like one's idea of Darnley, except as to the dress, which was superb, and in which he looked very handsome and dignified. Mr. Montgomerie, very good as "Pickwick :" his wife, "Mrs. Bardell," but a great deal too pretty for the character. Many of the ladies were in old Court dresses, many of the men in uniforms of various kinds. The affair was kept up with great spirit, and very late, and the guests seemed very well pleased.

Harry was dressed as a Cavalier, of about the time of the Restoration—a handsome dress, and he looked very well in it. Herbert, in his proper Artillery uniform. Arthur, in a sort of Spanish Bull-fighter's dress (made up for him in the house by Fanny and her maids), was extremely handsome.

Morewood wore the *new* Court dress : Mr. Anger- 1877.
stein his Deputy Lieutenant's uniform : William
Hoare was a very good sailor. Evelyn Bevan as
" Miss Angel " (after Miss Thackeray's novel),
looked very stately and even taller than she usually
does.

_____

Monday, January 22nd.

The *Conference* at Constantinople has broken up,
having accomplished nothing :—a complete failure.
The resolution or obstinacy and self-confidence of
the Turks, appears to have, for the time, baffled all
the Christian Powers, and has made them cut a
very poor figure. What was less to be expected,
Russia, which at the beginning of the business
appeared so threatening and formidable, now
appears to hesitate and hang back and to be
unwilling to push matters to extremity. Whether
this irresolution is real, and if so, what is the reason
of it ; whether Russia is really behind-hand with
her military preparations, or is kept in check by
want of money or by fear of Germany, time will
show.

In the meantime the Turkish Government have
granted to all their dominions a *Constitution* which I
suppose will be about as valuable as those of some
of the South American Republics.

_____

Tuesday, January 23rd.

Signed deed for sale of Vicarage Close at Mildenhall.

_____

Thursday, January 25th.

Dear Sarah and Albert Seymour, with their two
children arrived. Charlie now a little more than
two years old, is a beautiful child and very amusing.

Friday, January 26th.

Dear Arthur went back to school.

Two Miss Bevans and two Miss Bouveries came
to luncheon—charming music after it.

Amused myself with Gibbon's " Autobiography."

Tuesday, January 30th.

Lord and Lady Rayleigh arrived—also John
Hervey.

Wednesday, January 31st.

There was the most violent gale of wind yesterday
that we have experienced here for many years;
indeed, I hardly know that there has been any equal
to it since 1860. Several trees have been blown
down or broken off in and near the park. In
particular, a large picturesque, very old ash tree in
the park, of which Kingsley took notice when he
was here, has been broken off near the root, and
shattered to pieces in the fall : it was much
decayed. Also an elm opposite Phillips's house,
one in a paddock behind Scott's house, and others
of less consequence have fallen. But fortunately,
and strangely, neither of our grand old oaks in the
pleasure-grounds has suffered, nor indeed any tree

very near the house. Nor has any very serious
damage been done in the arboretum, though several
of the pines and firs have lost small branches.

Albert Seymour went away.

Mr. and Mrs. Martineau and Mr. and Mrs.
Sidgwick arrived.

Walked with Minnie to see the fallen ash tree.

Henry Wilson and his daughter and two of the
young Mr. Wilsons, and also the Thornhills dined
with us.

The weather of this month has been very remarkable for its mildness. In the whole course of the
month there have been only 13 days on which the
thermometer was down to 32 degrees or below it,
and the lowest temperature marked was 25 degrees.
There were 7 days on which it did not descend
below 40 degrees.

There has been no snow at all here. The grass
is beautifully green, and the garden-beds are already
gay with snowdrops (both the Crimean and the
common), violets, primroses, and here and there
anemones and crocuses.

Thursday, February 1st.

Showed the garden to Mrs. Martineau.

Saturday, February 3rd.

A very fine day.

Dr. and Mrs. and Miss Hooker arrived. We
walked through the garden with them.

1877.   Captain Horton and the Broke girls and the Percy Smiths dined with us.

We went to morning Church and received the Sacrament.

My 68th birthday. I have great reason to be very thankful to the Almighty that I have arrived at this age in fair health, and surrounded by so many blessings ; above all, such a wife, and such friends.

I see announced in *The Times* the death of Lady Smith, the widow of the great botanist, Sir James. If she had lived till May she would have ben 104 years old. She was really an extraordinary woman. She seems to have retained her faculties and the clearness and vigour of her mind to the last. Katharine Lyell has kept up a correspondence with her for several years past, and visited her at Lowestoft last year, when her mental powers were still wonderfully perfect.

LETTER.

Barton, Bury St Edmund's,
February 5th, 77.

My Dear Katharine,

I thank you heartily for your kind note and congratulations on my 68th birthday, also for the

intended present of your photograph, which I shall 1877.
be very glad to have. I am sorry to see in to-day's
Times the death of your friend Lady Smith, within
three months of the completion of her 104th year.
She seems to have preserved her faculties and the
freshness of her mind almost to the last. She was
a wonderful woman.

I am thankful to say that I am well in health,
though not feeling so strong as I did in the summer,
(but I never do feel as strong in winter as in
summer), and Fanny also is fairly well, though
working hard. I can well understand that you
must have a great deal of work and of care
and of disagreeables also, I am afraid, in
changing your house, and I can enter also into
your feelings of regret at leaving what has been
your home for so many years. But when the
change is completed, I hope you will feel that it is
for the better.

We have had a very pleasant party during the
last eight or ten days: and Minnie Napier, Sarah,
with her two little children, Dr. and Mrs. and Miss
Hooker and Edward are with us still. I cannot
write more just now, for I am a little tired, having
taken a walk with Dr. Hooker, and it is near post
time, but I will write again soon. With my best
love to Rosamond,

Believe me ever,
Your affectionate brother,
CHARLES J. F. BUNBURY.

JOURNAL.

1877. Very pleasant company in our house since the 25th of January, on which day dear Sarah and Albert Seymour, with their two little children, arrived. Dear Minnie came on the 29th. Lord and Lady Rayleigh, Mr. and Mrs. Sidgwick, Mr. and Mrs. Martineau, John Hervey, Joseph Hooker and Mrs. and Miss Hooker and my brother Edward made up the party—not at one time, but some coming and others going. Sarah looking very well, and as charming, good and loveable as ever. Mr. and Mrs. Sidgwick, who were new acquaintances,—pleasant. The John Martineaus we knew in some degree before, but have now much improved our acquaintance, and I like them particularly. *He*, uncommonly well informed, gentle and pleasing in manner, quiet, thoughtful, earnest ;—*she*, very lively and conversible, intelligent, with very pleasing manners. Their warm regard for the Kingsleys, and their friendship with Mr. Sanford, are additional links between them and us. Joseph Hooker not well,—in fact he complains of being overworked, and looks it. He has besides the regular work of the administration of Kew, which involves an enormous correspondence,—besides this, his co-operation with Bentham in the Genera Plantarum, his position as President of the Royal Society, which exposes him to incessant questions and applications from Government, and lastly, what he

says fatigues and *wears* him most of all, the late 1877.
hours of London society. Mrs. Hooker (a new
acquaintance for us, though she has long been a
friend of Katharine's), handsome, and looks clever :
she is quiet and gentle in manner. Hooker's
daughter is remarkably pretty, and has pleasing
manners.

Wednesday, February 7th.

Fanny went with me into Bury—I to the Jail
Committee—met there Rodwell, Barnardiston, Mr.
Young and Mr. Bence. Mr. Rodwell described to me
enormous damage done in his grounds at Ampton
by the storm of the 30th ; several fine cedars near
the house blown down or shattered,—one of the
largest of them torn completely up by the roots,
and so huge a mass of soil torn up with it, as to
look like a barricade : a very fine Catalpa; on the
lawn, completely shattered to pieces. We seem to
have escaped most fortunately.

Thursday, February 8th.

Read St. Luke's Gospel, chapter 6, in Greek,
with Alford's notes.
Finished reading volume 1 of Sir Erskine May's
" Constitutional History."

February 10th.

Opening of Parliament, on the 8th :—long and
spirited debates on the Address in both Houses, but

1877. as there was no division nor amendment even proposed, these can only be considered as a preliminary skirmish, or a sort of *reconnoisance en force*, to examine the adversaries positions. The Duke of Argyll was very vehement, not to say violent.

February 17th.

Mr. Pearson of Bury, a few days ago, gave us a fine *rough-legged buzzard*, which had been caught in a trap on the open country near Icklingham. It was not hurt, and is a very fine, handsome bird, full grown and in fine plumage. We are going to send it to the Zoological Gardens.

Joseph Hooker, when he was last here, noticed the beauty of a grass which we have in cultivation here, and on which the dry stalks and some relics of the seeding panicles of last summer remained ; and also of the large flowered Crimean snowdrop, which flourishes exceedingly here ; and asked me to send some of them to Kew. I sent him, accordingly, a good clump of the grass (which he tells me is *Piptatherum multiflorum* from the Mediterranean region) and some bulbs of the snowdrop, and also two young plants of the Himalayan horse-chestnut. This morning I have a note from him telling me of their safe arrival.

LETTER.

My dear Edward,

I am sure you will be sorry to hear of the 1877.
great loss we have suffered in the death of our
excellent servant Edgar Calthorpe. He was attacked with inflammation of the lungs, no longer
ago than Saturday the 10th, and the disease made
rapid progress; yet in the latter part of last week
there seemed to be a decided improvement, and
on last Saturday evening, Dr. Macnab spoke very
hopefully; but yesterday morning there was a
violent return of the fever, and he died between
9 and 10 o'clock last night. Dr. Macnab says that
the local disease in the lungs had quite subsided,
but that he had not strength enough to throw off
the fever which followed. In fact he seemed to
have been in a depressed and unhealthy state
for some time before, and he said himself that he
had never been well since his daughter died. We
are much grieved, for he is not merely a great loss
as a valuable servant, but he had been with us
a long time, and we had a great esteem and regard
for him and entire confidence in him. He was an
excellent man and a first-rate servant. We have
put off all the party of guests who were coming
to us—Mr. Courtenay and Lady Muriel Boyle, the
Walronds, the two John Herveys and others. We
are also with Minnie and her two dear little grand-

1877. children and Harry.   Fanny is moderately well, but much worn and harassed by anxieties and troubles (for besides Calthorpe, there has been an alarm about Mrs. Betts—Julia, whom you may remember as Fanny's young maid, who we took with us to Florence) ; and she has also had much vexation about her Aunt's will.   I am quite well, and I hope you have continued so since your return to London. Fanny says she is rather disappointed at not having heard from you since she wrote to you on hearing of the terrible death of Mr. Arthur Strutt.   What a dreadful accident that was !

Ever your affectionate brother,

CHARLES J. F. BUNBURY.

JOURNAL.

Saturday, Februrary 24th.

Mr. Bowyer arrived.

Fanny went to Bury to an examination of pupil teachers.   Archdeacon and Mrs. Chapman dined with us.

The two great debates, one in the House of Lords, the other in the Commons, have pretty nearly exhausted the Eastern question as far as our Parliament can deal with it ; at least they have left little of interest in any further discussion of it, until the affairs themselves are more clearly evolved. There have been some fine speeches on both sides, especially in the Lords, but on the whole the debates have been unsatisfactory, because too evidently leading to no result, and hardly even tending to one.

It is clear that the Opposition has no serious expectation of overthrowing, or of even materially damaging the Government, not clear that they even seriously wished to do it. In fact, the course pursued by the Government, since last September, or thereabouts, has approached so near to that which seems to be recommended by the Opposition, that not much solid ground seems to be left them; which rather spoils the interest of a debate. Lord Salisbury's speech was admirable—the best of all to my thinking; Duke of Argyll's also very fine; Mr. Gladstone's *second* speech very vigorous.

Tuesday, February 27th.

Looked over some of Sir Thomas Hanmer's MSS. Lady Host, Mr. Abraham and his daughters dined with us.

Mr. Henry Bowyer left us to day, having come on the 24th. He is an old friend, having paid us pretty regular annual visits for about 25 years past. He is a man extremely well informed on a great variety of subjects; a great reader and a great talker; talks very well too. He is a very amiable and very good man.

Saturday March 3rd.

Very mild and excessively damp.

Read J. S. Phillips's paper on " Farm Pests and

1877. Farm Friends." Wrote to congratulate him. I see
in the newspaper, the marriage of the famous Mrs.
Norton to Sir William Stirling Maxwell; which
Edward mentioned in his last letter to me as about
to take place, and as being the chief topic of talk at
the clubs. She is nearly 70, and is said to be in
very infirm health; the bridegroom was born in
1818, and is therefore more than ten years younger
than she is.

Dear Sarah and Albert Seymour arrived. Their
two children have been left with us since the 7th of
last month. Charlie, aged two years and four
months, is a superb child, and extremely amusing.

Wednesday, March, 7th.

Mrs. Horton and Miss Broke, and Mr. and
Mrs. Bland dined with us.

Read some of Arnold's Lectures on History.
Studied Baker on Liliacea.

Friday, March 9th.

Wrote to Sarah Sanford. Read part of Arnold's
Fourth Lecture on Modern History.

Clement arrived.

Sunday, March 11th.

Rose Kingsley with us.

Read St. Luke, chapter xii., verse 22 to 40,
in Greek. Mr. Courtenay Boyle and his wife Lady
Muriel; Rose Kingsley; Lady Hoste; Mr. and Mrs.

Holden ; Agnes Wilson ; Miss Maitland ; John and 1877.
James Hervey, and Mr. Charles Strutt ; besides
Minnie and my nephew Harry.

———

Monday, March 12th.

Dear Sarah and Albert Seymour, with their
children, went away.

Poor Fanny's pet parroquet died.

Lady Muriel Boyle is a daughter of Lord
Cawdor, a very pleasing, intelligent, interesting
person. Her talk about her own country (South
Wales),—the beautiful scenery about her father's
estates in Pembroke and Caermarthenshire, the
famous Golden Grove, so associated with the name
of Jeremy Taylor, is one of his residences, was very
interesting to me ; she has a great taste for botany,
and seemingly, good habits of observation. Mr.
Courtenay Boyle (our Poor-law Inspector), is a
clever, well-informed, agreeable man. These two
are new acquaintances, and very pleasant ones.
The rest were old acquaintances. Mrs Holden was
Alice Praed, a very nice girl.

Rose Kingsley is as animated, warm-hearted,
eager and interesting as ever.

———

LETTER.

Barton, Bury St. Edmund's,
March 12th, '77.

My Dear Katharine,

I am glad that you have at last accomplished the move into your new habitation, and I

1877. hope that before long you will begin to feel yourself
somewhat at home in it, though I can well
sympathize with your feelings of regret in leaving a
house which has been your home for so many
years,—where your children have grown up, and
where you have experienced so many joys and
sorrows. It must necessarily take some time to
reconcile one to such a change.—But I hope and
trust that you will find your new house healthy and
comfortable, and I am very glad that when we go to
London, we shall find you so much nearer. I hope
there is good room in your new house for your
botanical collections and your books. Have you
unpacked your herbarium yet ? My chief botanical
employment lately has been studying Mr. Baker's
papers on Liliaceæ in the latter volumes of the
Linnean Journal and arranging my plants of that
family accordingly. I see, that in the last number of
the *Journal*, there is a re-arrangement, by Bentham,
of the whole of the Monocotyledons, which will be
important to study. I have got Woodward's new
book on "The Geology of England and Wales,
which seems likely to be very useful, especially
when one travels.

Dear Sarah Seymour and her husband and
children left us this morning, and I was very sorry
to part with them. Their children have been here
more than a month ; Charlie is a superb little fellow,
and excessively amusing. I shall be very glad when
we go to London, to renew my acquaintance with
my other godson, Charlie. Dear Rose Kingsley is
with us now, as also Minnie Napier, Clement and

Harry. My nephew George has had a narrow 1877. escape from a dangerous illness; he caught a fever, something of a malarious sort, while travelling in Corsica, and with this fever on him he went to Rome and Naples, and went about sight-seeing.

Most happily the illness did not come to its height till he had returned to his mother at Mentone; there he was extremely ill for a few days, but got well through it, and his mother in her last letter reported him as quite recovered.

We had a pleasant party in our house all last week. Mr. Courtenay Boyle (the Poor-law Inspector of our district), and his wife Lady Muriel are new acquaintances, very pleasant ones; she, very pleasing, gentle, intelligent and interesting, with a strong taste for botany and for scenery; he, a very clever, well-informed and agreeable man. The rest of our party were old acquaintances.

I am reading some of Burke's speeches, almost the only modern speeches I know which are interesting to read when the special interest of their subjects has passed away.

Poor Fanny has lost her pet parroquet (one of the grey kind, with yellow cheeks and a crest); a great pity, for it was the tamest, most loving, most caressing little creature possible, at least I never saw a bird like it in these qualities except poor Zoë, which you may perhaps remember. It has died, as it seemed, of a gradual decline.

1877. *(March* 13*th).*   I am very glad to hear of Frank's safe return from America.

Believe me (with much love to Rosamond),

Your very affectionate brother,

CHARLES J. F. BUNBURY.

JOURNAL.

Friday, March 16th.

Business with Scott.   Preparations for the audit. Mr. Claughton and the Percy Smiths came to luncheon.   Read aloud, in the evening, the first act of " Twelfth Night."

Rose Kingsley left us, to my very great regret,— she could no longer remain away from her mother.

Saturday, March 17th.

My Barton rent audit—very satisfactory.

Luncheon afterwards to my tenants, Mr. Percy Smith, and my nephews.

In the evening read aloud 2nd Act of " Twelfth Night."

March 20th.

Examined the flowers of Triteleia uniflora (which we had received under that name from Mr. Fish), identified it with Milla uniflora of the *Botanical Magazine*, and made a note.

Wednesday, March 21st.

Troublesome business about the Mildenhall Schoolboard. Talk with Scott and Betts.

A visit from Mrs. Abraham. In evening, read aloud—1st act of "As You Like It."

———

Saturday, March 24th.

A very fine day, quite mild.

Sent a subscription to the fund for relief of families of fishermen lost at sea. Finished reading aloud in evening—" As You Like It."

———

Wednesday, March 28th.

The Bishop of Ely and his Chaplain, Mr. St. John, Mrs. Douglas Galton and her daughter and Archdeacon and Mrs. Chapman arrived. Mr. Percy Smith dined with us.

———

Thursday, March 29th.

The Bishop and his Chaplain went away—also the Chapmans. Read St. Luke, chapter 13 to verse 21.

Cecil arrived.

Lady Hoste and the Holdens dined with us.

———

Friday, March 30th.

A beautiful day. Read prayers with Fanny.

Had a pleasant little walk with Minnie, Miss Galton and Cecil.

———

Saturday, March 31st.

Lady Rayleigh and her sons Edward and Headly, arrived; also my nephew George. Went on with Sir Ralph Abercromby.

Dear Minnie left us, to my very great regret. She is an invaluable friend and companion, quite a sister.

More severe weather, and much more snow in this month and the last days of February than in all the previous part of the winter. On the 1st and 2nd, the lowest of the thermometer respectively 20° and 24°; the 3rd and 4th, much warmer.

Monday April 1st

I showed Lady Rayleigh the gardens and arboretum. Lord John Hervey arrived. Mrs. Horton and Freda dined with us.

Tuesday, April 3rd.

Colonel du Cane, Inspector General of Prisons, arrived. George went away. Arthur returned from Haileybury on account of scarlet fever in the school.

Thursday, April 5th.

Began to read Froude's "Short Studies on Great Subjects," third series.

The Bishop of Ely, with his Chaplain, Mr. St. John, Archdeacon and Mrs. Chapman and Mrs. Galton, and her daughter Gwendoline arrived on the 28th. The Bishop, who is a remarkably

agreeable man, highly accomplished, and of charming manners, came for a Confirmation, and could
not stay beyond the next day, the 29th ; and the
Archdeacon (also a very agreeable man) also went
away on that day. Since then we have had for
guests—Lady Rayleigh* and two of her sons, Lord
John Hervey, Colonel du Cane, and my nephews
Cecil and George. Now all are gone except Mrs.
and Miss Galton.

Yesterday, the 4th, in the latter part of the
afternoon, there was a very violent thunderstorm,
the first (as far as I know) of this year. It began
here about 5.30 p.m. The lightning was very
vivid, and seemed near. Just before the storm
broke the colouring and appearance of the heavy
masses of clouds, which came rolling rapidly up
from the West, were strangely lurid and almost
awful.

April 7th.

Received a charming letter from Rose Kingsley.
Among other things, she tells me that her mother
has just received the first copy of the fifth edition of
her book, "The Life of Mr. Kingsley." This is
pleasant news.

April 9th.

Saw a *Jay* hopping and feeding on the lawn ;—
the first time that I remember to have seen one on
the ground: indeed the first time that I remember

* The Dowager.

1877. to have had so good a view of one of these beautiful birds in a state of nature.

_____

Tuesday, April 10th.

Mrs. and Miss Galton left us. Their visit has been very pleasant. Mrs. Galton is very clever, very well informed and remarkably agreeable, her conversation very brilliant. Gwendoline is a pretty girl, very lively and light-hearted, good-humoured and good-natured.

_____

Wednesday, April 11th.

The 33rd anniversary of our happy engagement.

=======

LETTER.

Barton Hall, Bury St. Edmund's.
April 18th, 77.

My Dear Katharine,

We think of going up to town on the 24th, next Tuesday, if the weather is tolerable, and if Fanny is well enough.

We have been quite alone (except Arthur, who is very flourishing), since Mrs. and Miss Galton left us. Their visit was exceedingly pleasant.

Harry is gone back to Cambridge.

Fanny is excessively busy about a thousand things, in spite of her cough, which does not abate her energy, though it confines her to a few rooms.

I can enter into your regrets at being so long debarred from the use of your herbarium.

I have lately been looking through part of the collection of Simla mosses which you gave me long ago, but have by no means arranged them yet. 1877. There seem to be several interesting forms, but it strikes me as curious, that (as far as I have gone yet), I do not recognize any of the Nepaul mosses figured in the Musci Exotici. There is an important paper on Indian mosses by Mr. Mitten in the *Linn. Soc. Journal*, but I am afraid my eyes will not be equal to making out his excessively minute characters.

I should like very much to see Miss North's paintings, of which I have heard wonders, from Edward, as well as from Mrs. MacMurdo, and now from you, and I hope I shall one day have that treat.

There is in the April number of *The Edinburgh*, a good article on Charles Kingsley's Life. There is also an article on South Africa, which is directly and avowedly a reply to one on the same subject (said to be by Mr. Froude), in *The Quarterly* of January : they are written from opposite points of view ; *The Quarterly*, entirely from the Dutch-colonists' point : *The Edinburgh*, from that of the former Colonial Office, or partly of the Aborigines Protection Society. *The Quarterly* article is much the best written, but the writer evidently thinks that the natives (as well as the Irish Papists), had better be "improved off the face of the earth").

I am very glad to hear a good account of Charlie, and shall be very glad to renew my acquaintance with him.

With Fanny's best love, believe me ever,

Your affectionate brother,

CHARLES J. F. BUNBURY.

JOURNAL.

1877.   Found cones of Ab. Menziesii.

A beautiful day—real spring.

Fanny went out, first time for many days. I enjoyed a walk in the garden with her.

Excessively wet and cold.

A visit from Lady Hoste—very pleasant.

Arrangements in my museum.

Nothing can be more gloomy and threatening than the aspect of Eastern affairs; there seems hardly a hope that war can be averted, at any rate between Russia and Turkey: and, war once begun, God only knows how far it may spread.

The extraordinary rescue of five miners, in a coal mine at Pont-y-Pridd in the Rhondda valley in Glamorganshire,—imprisoned by a sudden rush of water into the mine—living without food in that horrible situation for *nine* days, and at last brought out alive and likely to recover.—It is marvellous. The fortitude and patience of the imprisoned men, and the courage of those who laboured (at the imminent risk of their own lives), to release them, were admirable.

Very fine.

A pleasant stroll with Fanny in the garden.

Business with Scott—Beck Row Farm, &c.

Tuesday, April 24th.

A pleasant drive with Fanny in the pony carriage.

Thursday, April 26th.

Weather bitterly cold.

Up to London with Arthur and six servants—arrived safe—thank God.

Arrived at 48, Eaton Place, about 4.

The war has actually begun; the Russians have not only crossed the Pruth and entered Roumania, but have at the same time crossed the frontier at the Asiatic end of the Black Sea, from their Caucasian territories, and a skirmish is reported to have taken place near Batoum. So begins a war which is likely to be an awful one, and awful is the responsibility of those who have brought it on.

Friday, April 27th.

A cruel, cold east wind.

MacMurdo came to luncheon—very pleasant.

Drove out by myself—had a glimpse of Norah Aberdare: called on J. C. Moore, whom I afterwards saw at the Athenæum.

Montagu, Susan, and Mimi MacMurdo and Edward dined with us—extremely pleasant.

The reports, from the two sides, of the first collision between the armies (near Batoum), gives us a pretty sample of the degree of *certainty* and *credibility* which we have to expect. According to the Turkish statement, the Russians lost 800 men ; according to that of the Russians, their loss amounted to one Cossack killed and a few wounded.

————

April 29th.

Montagu MacMurdo tells me, that the soldiers of the Turkish army are undoubtedly good, but their officers are bad : and if (as one report says), the Turks will not consent to employ foreign officers in their army, they will suffer for their pride. He says, it is certain that the Turkish navy is at present decidedly superior to the Russian ; and they have not only great and powerful iron-clads, but a strong flotilla of steam gun-boats on the Danube, which will make it difficult for the enemy to cross. He believes that, since the war of 1828-9, the two nations have made somewhat corresponding advances in military organization and efficiency ; but in those days Russia had by her fleet the command of the Black Sea: now Turkey has it, which makes an important difference.

————

Monday, April 30th.

MacMurdo has pointed out to me the advantage

of the position of Galatz, which the Russians have 1877. seized at the very beginning of the war; how com- manding the great bend of the Danube, where that river flows first to the north and then to the east. It covers their flank against any operation of the Russians from the coast; and also at the same time, secures for them a good base of operations on the river. He says that if the Russians could get possession of Widin, the road from thence to Adrianople is comparatively open and easy, *turning* the Balkan range by its western extremity. But the line from Bessarabia to Widin would be danger- ous if they were not secure of the friendship of Austria.

———

Tuesday, May 1st.

Fanny drove out, first time since we came to London. We visited Miss Phillips, and I by myself visited Minnie Powys and Mrs. Young. Both very agreeable.

We have been shocked and grieved by the news, this morning, of the death of the wife of George Hervey, the Bishop's second son. She died in child-bed, on the 29th, having brought forth a daughter the day before. They had been married less than a year. I never saw her, but she is said to have been a charming person, and the marriage gave great satisfaction to all his family. This death will be a painful shock to Sarah and Kate, as well as to Lady Arthur and the Bishop.

———

A visit from Mr. Robeson to talk over business of
Mildenhall Vicarage.

Yesterday I went with Susan MacMurdo to Miss
North's, and spent two hours in looking over her
paintings of flowers, which are really wonderful.
They are all done in oils.   I certainly have never
seen any flower paintings that at all approached
them, for they have all the greatness and artistic
effect of the finest Dutch paintings of such subjects,
with truth and accuracy of detail which would
satisfy any botanist.   Some works of Ferdinand
Bauer, which I saw formerly at Robert Brown's,
could best bare a comparison with them.

Dr. Hooker has seen Miss North's pictures, and,
as I understand, admired them as much as I do.

Miss North has been an extraordinary traveller,
as well as an artist ; she seems to have visited most
of the tropical regions, which are rich in vegetation,
and everywhere made studies, indefatigably, from
the living plants in the forests.

I was incorrect in saying that she was a painter of
flowers, her paintings represent, not merely flowers,
but fruits, trunks of remarkable structure, groups of
epiphites in their natural situations on old mossy
trees or rocks, Palm trees and Fern trees,
grand specimens of Banyan and India-rubber
trees with their extraordinary air-roots and multi-
plicity of stems, in short, all the beauties and
wonders of the vegetable world.   The paintings
which I saw, were done in Java, Borneo (Sarawak),
Rio de Janeiro and Teneriffe.   Those of the two

latter places were particularly interesting to me, but 1877.
I am not sure that the oriental ones are not the
most admirable.

Some of these works which particularly interested
me were :—

The Mangosteen, with flowers and fruit, and
sections of the same—agreeing perfectly with the
figure in the *Botanical Magazine*.

Cinchona, one of the species cultivated in Java,—
a panicle in full blossom, and another in seed, with
leaves in the rich green of their perfection, and
others faded to a splendid crimson—a beautiful
subject, and exquisitely painted. Nepenthes—
several species, one specially magnificent (Raffle-
siana ?) with its various forms and rich colours
of pitchers on an old tree.

The Durian — that strange fruit described by
Wallace—whole and broken open.

Mangrove trees (Rhizophora) branches with
leaves, flowers, and fruit in their natural situation,
hanging over a " Mangrove swamp," the tangled
stems and strange roots of the trees forming a back-
ground to the botanically finished picture. I never
before saw a picture which gave me a distinct and
lively idea of the famous Mangrove.

The Talipot Palm of Ceylon, in its various stages
of growth.

The Bread Fruit.

The Areca Palm in various stages of growth,—
most beautifully painted ; it is marvellous with what
truth and grace she has represented the character of
the leaves of one family of Palms—the light, pliant,

1877. quivering, grass-like character of such as the Arecas
and Euterpes. The Dragon Tree in Mr. Smith's
garden at Orotava, which I well remember—admirably true to nature.

_____

Friday, May 4th.

Visits from Mr. Hall and Mrs. Bowyer, with two
of her children.

Saw Aberdare at the Athenæum.

Minnie Powys, Mrs. and Miss Ellice, Edward
and Clement dined with us—very pleasant.

Miss Jackson (the Bishop of London's daughter),
whom we met at dinner yesterday at the Bishop
of Ely's, told me, that three of their fine old trees
in Fulham Palace Gardens—of those trees planted
by Bishop Compton—were destroyed by a gale
last January; and their famous Black Walnut,
the finest in this country, though not blown down,
lost a large limb at the same time.

_____

Saturday, May 5th.

Dear Arthur went to a private tutor at St.
Leonards,—Mr. Adams.

We drove out late in afternoon—saw Fanny
Jeune and the Clements Markhams and Captain
Markham.

Sarah and Albert and Edward dined with us—
very pleasant.

Calling on the Clements Markhams, we met Captain
Markham, the celebrated Arctic voyager, a man of
pleasant appearance and manners. We saw his
famous and fine large black dog " Nellie," which

was his constant companion all through the Arctic 1877.
voyage—a very handsome and good-natured animal.
Clements Markham showed us in a glass case
several characteristic animals of the Polar regions :
those curious little beasts the lemmings (which are
smaller than I had imagined from the prints) : the
Arctic fox in its winter dress, a very pretty animal :
the snowy owl, a very handsome bird, the Iceland
falcon : the beautiful ivory gull in its pure white
plumage : also the eggs of the eider duck, bedded in
the down from its nest, and the head and skin of a
musk ox (or musk sheep as Huxley prefers to
call it).

Monday, May 7th.

Dear Minnie arrived.

Walked with Fanny : we visited Mrs. Prideaux
Brune and Mrs. Rumley.

Tuesday, May 8th.

Visits from dear charming little Charlie Seymour.
Wrote to Rose Kingsley.

We dined yesterday with the MacMurdos at
Fulham ; Lady Head and Edward went with us :
we met Miss North, Mr. and Mrs. Coltman, Mrs.
and Miss Talbot, McLeod of McLeod, Mr. Card-
well. It was a pleasant evening. The Mac Murdos'
house and conservatory are beautiful ; Susan has
certainly a remarkable gift of taste. The two eldest
girls, Emily *(Mimi)* and Susan, are quite lovely.
Miss North brought a large number of her exquisite

1877. botanical paintings which I had great delight in looking over and discussing.

Miss North is a very gentle, quiet, refined person, far from masculine, though she has travelled quite alone through so many wild countries.

A visit from Sir Frederick and Lady Grey.

The Louis Mallets, Douglas Galton, and Mrs. and Miss Galton dined with us—a pleasant party.

We dined yesterday with Katharine, at her new house, 9, Cornwall Gardens. It was an exceedingly pleasant small party; Mr. and Mrs. Maskelyne, Miss Kinloch, Lady Head and Amabel, George Bentham, Leonard and Mary Lyell, Edward and a Mr. Grey. I sat by Mrs. Maskelyne, who is a remarkably pleasing, well-informed, interesting person; and afterwards I had much talk with Miss Kinloch, who is also extremely pleasing.

Katharine's new house (in the far south-west, pleasantly situated—the rooms good (at least those on the " drawing room floor,") and very gracefully and suitably furnished and decorated.

Mr. and Lady Barbara and Miss Yeatman came to luncheon.

A visit from John Moore. Sarah and Albert dined with us, and then went with Minnie to the play.

Saturday, May, 12th.

We three (Fanny, Minnie, and I) went down to St. Leonards by the 12.30 train, had a good journey —Arthur met us on the platform at St. Leonards, and went with us to the Victoria Hotel.

———

Sunday, May 13th.

We all three went to Church ; first time Fanny had been at Church since March. Poor Arthur attacked with ophthalmia.

———

Monday, May 14th.

At Victoria hotel, St. Leonards.

Fanny went to spend the afternoon with Arthur.

Minnie and I had a pleasant drive through a pretty country to Bexhill, up to the Church and back.

Received a very pleasant letter from Lady Louisa Legge, from Rome.

———

Tuesday, May 15th.

St. Leonards.

Fanny spent the day with Arthur, as before.

Minnie and I drove to Battle—saw the Abbey in an unsatisfactory way, but the drive through the pretty country was very pleasant.

———

Wednesday, May 16th.

We returned to London.

———

Thursday, May 17th.

48, Eaton Place.

1877.   Went out with Fanny.   We called on Kate
Hoare, and saw her pretty baby.

LETTER.

48, Eaton Place,
May 17th. 1877.

My Dear Lady Louisa,

I thank yon heartily for your charming
letter from Rome, as well as for your photograph,
which I admire very much indeed ; it is both an
excellent likeness and extremely pleasing, and I
value it highly.

It was very kind of you to think of sending it to
me.   I received your letter at St. Leonards, whither
we had gone for a few days.

I had heard and was much grieved to hear, that
you had lately been ill at Rome, so that your
agreeable letter was doubly welcome as shewing me
that you were in pretty good health.   I am very
sorry that there is no chance of our seeing you in
England this year, but I cannot wonder when there
is no pressing reason for returning ; there is such a
charm in Italy, and especially in Rome, that I
cannot wonder at any person of good taste liking to
remain there.   I trust, however, that you will not
stay so long at Rome, as to risk being caught by the
seriously unhealthy season.

There is certainly a curious charm about Italy.
I often dwell with delight on my recollections of it,
and I think it is the only foreign country that I feel

a longing to see again; but I do not think we shall 1877.
be able to manage a visit next winter.

I am very much entertained by your account of
the proceedings at Rome, especially of the way in
which they transform houses. It seems the next
thing to what I have read of as happening in
America, where (it seems) they think nothing of
setting a house on rollers, and rolling it to the other
end of a town, without disturbing the music or other
occupations of the inmates.

It must have been amusing to see the Japanese
Ambassador and Ambassadress.

Have you seen that interesting *prisoner*, the Pope?
He seems likely to surpass the other Princes of
Europe, in length of days at any rate, which speaks
well for the healthiness of Rome.

I feel that I can make but a poor return for your
delightful letter, that in fact I have but little to tell.
We have been but a short time in London, and
between the bad weather and Lady Bunbury's ill-
health until we went to St Leonards, we have seen
little or nothing; we have, indeed, seen several
friends, whom it is a great pleasure to see, among
others, Lady Barbara and Mr. Yeatman; but
what I meant is, that we have seen no exhibitions
or sights. I have not yet attempted to make my
way into the Royal Academy, and have not yet seen
the Grosvenor Gallery, though I have heard several
opinions about it; all seem to agree that the Gallery
itself is beautiful, but there are various opinions
about the pictures.

But you who are living in the very head-quarters of

1877. painting and of all the fine arts, will with reason, care little for pictures of the present day.

I have read nothing very interesting (at least nothing new) since the Life of my friend, Mr. Kingsley, which indeed, is no longer quite a novelty, and which interested me particularly, as I had a very great regard for him; it is admirably well done by his widow. I took advantage of a slight pause in our supply of new books, to return to some of the old ones, and in particular, to read over several of Burke's admirable speeches. What a great man he was! superior to my thinking to either Pitt or Fox.

The war is almost too great, too awful a subject to touch upon. I have a horror of it, and wish for nothing but peace. I cannot heartily wish success to either party. I am utterly against the Turks, but not strongly for the Russians, because I cannot forget their former doings in Poland and elsewhere. I remember you used to be rather an admirer of Russia; do you side with it now? I most earnestly wish that we may not be drawn into war on either side, but I am very much afraid lest Lord Beaconsfield's crooked policy should entangle us, before we are aware.

Lady Bunbury sends you her love, and is very sorry that we must not hope to see you in England this year.

Believe me, Dear Lady Louisa,
Yours very sincerely,
CHARLES J. F. BUNBURY.

JOURNAL.

Miss Adeane and Mr. Martineau came to luncheon.

We dined with Mr. and Lady Muriel Boyle,— Met Mr. and Mrs. Lestrange, Miss Vivian and Mr. Brand—pleasant.

We returned, the day before yesterday, from St. Leonards, where we had spent three days and parts of two others (from the 12th to the 16th); having gone thither partly for change of air for Fanny, who has been for several weeks in an uncomfortable, depressed state of health, the result of an influenza cold, in March,—and partly that we might see Arthur, whom we have sent to read with a clergyman there, who takes pupils, Mr. Adams.

Our trip answered perfectly; Fanny has returned in much better health, and we found Arthur very happy, and received a very favourable impression of the family in which he is settled. Fanny was very much pleased with what she saw of the arrangements of Mr. Adams' establishment, and especially with his mother, Mrs. Lees and his two half-sisters, who manage it. I saw only Mrs. Lees, a pleasing old lady, who seems simple, kind, motherly.

We enjoyed very fine weather during most part of the time we were at St. Leonards, and had some very pleasant drives in the pretty country near

1877. it. We (that is Minnie Napier and I, for Fanny preferred staying with Arthur), drove on the 14th to Bexhill, and on the 15th to Battle. The country, especially in the way to Battle, is decidedly pretty and pleasant, considerably undulated and varied in surface, with much copse-wood and rich cultivation. I was much interested in the drive to Battle, which enabled me to understand, much better than before, the nature of the country which was the scene of the battle of Hastings and of the preliminary movements. The broad ridge of high land along which the Norman army marched from Hastings is very well marked, commanding a wide view of the lower country on each side ; so is the depression from which the ground rises again to the height now occupied by the town and abbey of Battle. The remains of the Abbey do not appear to be very striking, and we did not see it satisfactorily, being hurried through by a conceited pragmatical fellow, who repeated a lesson like a parrot. I gained however a tolerable idea of the situation of the Abbey, and therefore of the field of battle.

All this country being in its full beauty of spring verdure and spring flowers, afforded delightful scenes : the various delicate tints of the young leaves in the copses and hedges, the brilliant grass, the blossom of the fruit trees, the hedge-banks spangled with primroses, the copses blue with the wild hyacinth.

Called on Lady Lilford (the Dowager), and had a pleasant talk with her.

Dear Kate and Charles Hoare, Minnie Powys, Leonard Lyell, and Clement dined with us : a very pleasant little party.

———

Monday, May 21st.

Wrote to Arthur.

Admiral Spencer, Susan and Mimi MacMurdo came to luncheon.

Visited George Bentham.

———

Tuesday, May 22nd

Clement and Willie dined with us. A visit from Louis Mallet. Talked of the extraordinary proceedings at Paris—MacMahon's *coup d'etat*. It has put a stop for the present, and, it is feared altogether, to the negotiations for a new commercial treaty in which Mallet was specially interested. He thinks it most disastrous in every point of view. It cannot have even a temporary success, unless the Government are able, by means of corruption, to "manipulate" the elections, so as to gain the appearance of a majority. The result must be bad either way ; whether it be the temporary triumph of the clerical party, which would give great offence and alarm to Italy ; or a strong Republican reaction, which would irritate Germany. Mallet thinks that MacMahon is a very honest man, but narrow-minded, and with a military rigidity in his notions.

1877.   If the Government are decisively beaten in the
elections, MacMahon will probably resign, and in
that case General Chanzy is the man whom Mallet
has heard spoken of as his probable successor.

LETTER.

48, Eaton Place, S.W.
May 22nd, 1887.

My dearest Cissy,

Thanks very much for your kind letter
of April 30 from Pisa, and of May 10th from
Florence, and especially for the full information
which they give concerning my dear mother's tomb
at Genoa. I thank you very much for the trouble
you took in going to see it, and to ascertain all
you could about its condition. I am very sorry
to hear it is in such a dilapidated state. I will
communicate with Mr. Brown on the subject.

Harry is, I hope and believe, reading well at
Cambridge.

We have seen many pleasant friends since we
came to town, especially dear Kate Hoare, Lady
Lilford (the Dowager), Minnie Powys, the Louis
Mallets, Mrs. Douglas Galton, the Clements Markhams, and of course Sarah and Albert, and the
MacMurdos often : we had two very agreeable
dinner parties at the Macs, and at Katharine's, and
some nice little parties here. But we have seen no
exhibitions or *sights* yet. We have however seen a
large collection of the most beautiful botanical

paintings by far, that I ever set eyes on, done by 1877.
a lady, Miss North, a friend of Susan MacMurdos,
who has travelled alone through a great many of the
finest tropical countries, and made all these pictures
on the spot, in the forests and swamps. They
are wonderful. I have not time to enter either on
the subject of the War or that of the extraordinary
proceedings in France. I hope we shall soon hear
a good account of yourself and dear Emmie, to
whom give my best love ; and I hope you will find
a healthy and pleasant abode for the summer. I
think you are quite right in not returning to
England this year.

Believe me ever,

Your affectionate brother,

CHARLES J. F. BUNBURY.

JOURNAL.

Wednesday, May 23rd.

We dined with Lord Talbot de Malahide—met
Lady Vaux, Miss Mostyn, the Gambier Parrys,
Miss Blackburne.

Lady Rayleigh came to luncheon, also Katharine
and Rosamond.

Thursday, May 24th.

Dr. Marcet, talking of MacMahon's *coup d'etat*,
remarked the singular unanimity with which the
English newspapers of all shades of politics, as well

1877. as those of Germany, Italy, and other Continental countries. have condemned or deplored this extraordinary measure. The papers devoted to the Pope have been, I fancy, the only ones which have approved it.

Our dinner party:—Lady Charlotte and Lady Octavia Legge, their brother Colonel Legge, Sarah and Albert, Charles Hoare, Leonard Lyell; afterwards in the evening, Kate Hoare, Caroline Hervey, Agnes Kinloch, Mary Lyell, Frank Lyell, the Hortons and Freda Broke—all very pleasant.

Friday, May 25th.

A visit from dear little Charlie Seymour.

We went to see Leonard and Mary Lyell, at Onslow Gardens, and especially to see their dear little boy, Charlie, one of my many Godsons. He is a charming little fellow; not such a superb child as Charlie Seymour, but very pretty, very intelligent and interesting, indeed rather precocious, so that there may be occasion for some caution lest his brain should be over-excited.

Went to see the Grosvenor Gallery in Bond Street. The Gallery itself is very beautiful; the pictures, several very good, and many very strange.

Monday, May, 28th.

Our dinner party:—Lord and Lady Rayleigh, Sarah,* Mr. and Lady Barbara Yeatman, Wil-

---

* Albert was unfortunately kept away by military orders.

loughby and Miss Burrell, Mr. and Mrs. Lecky, 1877.
Baroness (Mrs. Lecky's sister), Sir Francis Doyle,
Herbert Praed, Katharine, Rosamond, and Frank
Lyell, Edward and Clement. Very agreeable.

Mr. Lecky told me (what I am sorry to hear)
that Mr. Motley is in very bad health, and has
been ordered by his doctors to abstain from writing
and all intellectual exertion ; so that there is little
hope of his being able to continue his history.

Willoughby Burrell's wife (the daughter, I understand, of a physician at Dublin), is a remarkably
agreeable and interesting young woman, lively,
clever, and very well read ; quite an acquisition to
our circle of acquaintance. Mrs. Lecky, a Dutch
lady, is likewise very agreeable. I sat between
these two ladies at dinner, and found myself
very well placed.

Wednesday, May 30th.

The 33rd anniversary of our happy wedding, God
be thanked.

We dined with the Douglas Galtons—met Lord
and Lady Lovelace, General Laffan (Governor of
Bermuda), Lady Strangford, the Gambier Parrys.

Thursday, May 31st.

Fanny and Minnie went to the MacMurdo's
afternoon party at Rose Bank. I stayed at home.

I see in the newspapers, the death of Mr. Motley,
the great historian. It was only three days ago

1877. that I learned from Mr. Lecky that he was in very bad health. He is a great loss. Who now will take up the history of the thirty years war?

General Laffan, whom we met at the Douglas Galton's, is of opinion that the Turkish army is very inferior in all respects to the Russian, and has no chance against it in the field. The Russians he thinks, may meet with a good deal of difficulty and delay in crossing the Danube, and may also encounter considerable difficulty, as to food, in crossing the Balkan, where no supplies can be obtained, and it may be necessary to carry several days provisions for the army. But in the plains, he does not think that there will be any serious difficulty as to provisions and he does not believe that any formidable resistance can be made by the Turkish army.

The weather all through this month of May has been more persistently disagreeable, cold and wintry, than in almost any other I remember.

—————

Friday. June 1st.

Our dinner party :—Lady Wilhelmina Brooke, Lady Octavia Legge, Captain Legge, the Palmer Morewoods, the two eldest Miss Egertons, the Leonard Lyells, Agnes Kinloch, Mimi MacMurdo Harry Bruce, Captain Medlicott, Clement.

A charming party of ladies.

—————

Saturday, June 2nd.

Mimi MacMurdo stayed with us till the afternoon

when her mother came for her. William Napier 1877. and John Herbert came to luncheon. Fanny and Mimi went to a large evening party at Hertford house.

Joseph Hooker has been made a Knight Commander of the Star of India. I am very glad, though it is a tardy recognition of his merits. In any other country of Europe, he would long since have been made at least a baron.

Katharine tells me that Lord Salisbury accompanied the communication with a very satisfactory letter to Hooker, showing a just appreciation of his merits.

Sunday. June 3rd.

A beautiful summer day. Read prayers with Fanny.

Visited the Egertons, very pleasant—and General Romley.

Monday, June 4th.

A beautiful summer day, quite warm.

Mrs. Mills and Admiral Spencer came to luncheon. Visited Minnie Powys, and met there Wilhelmina Brooke and Octavia Legge.

Looked into the Zoological Gardens. The snakes numerous and tolerably alert; noticed the *Ringhals* snake, " Sepedon," (or *Naia*) *haem achates*, from the Cape, of which species I killed one on the Cape Flatt, in January, 1849, but have never seen a specimen since ; also several rattlesnakes, horrible

1877. looking creatures; the famous Russell's viper, from India, finely variegated, but almost as repulsive, in the look of its head, as the rattle-snake; and some beautiful pythons.

Admired the tigers, of which there are several, now seen to great advantage in their new, large, out-door cages, in which they have ample space to move about, and enjoy the fresh air and light.

These out-door cages or enclosures are at the back of the new " Lion house," from which they open. They are securely iron-barred at the sides and above, but open to the air and light.

Hornbills, several, very curious and remarkable birds, very lively and thriving.

Burchell's Bustard, a splendid bird of that family, with plumage very finely mottled and pencilled—the same of which the head is given in Burchell's Travels.

––––––

Tuesday, June 5th.

Our dinner party.—Lady Mary Egerton and one of her daughters, Lord Talbot de Malahide and Miss Talbot, Mrs. Mills, Admiral Spencer, Sir Alfred Horsford, the MacMurdos, the Gambier Parrys, Colonel and Mrs. Lynedoch Gardiner, the Clements Markhams, Mr. Jeune, and Fanny. Several more in the evening.

––––––

Wednesday, June 6th.

A very cold disagreeable day, ending in rain.

Harry Bruce and William Hoare came to luncheon.

We dined with Sir Edward and Lady Blackett, 1877. in Portman Square—met Lady Eastlake, Mr. and Mrs. Reeve, Lord Justice Sir W. James, Mr. W. Spring Rice, &c.

A fine day.

We (including Minnie) dined with the Morewoods, met Miss Rodney, Mr. and Mrs. Paget, Mr. and Mrs. Allsop, Constantine Hervey, Captain Medlycott, Sir Charles and Lady Blois, and others.

My first visit, this year, to the Royal Academy; only a short and hasty visit, to take a rapid survey of some of the rooms, and note a few pictures which struck me. I was much pleased with some, delighted with *Millais'* " The Sound of Many. Waters," a rapid headlong river, broken by rocks, and flowing through woods, such as one may see in the Highlands, or in North Wales. The painting of the rocks especially in the foreground, is most admirable ; so is the colouring of the woods, just touched by autumn.

" Arundel," by *Vicat Cole*, another superb landscape. *Long's* " Egyptian Feast," very clever and entertaining, but it does not strike me so much as his " Babylonian Marriage Market," two or three years ago. The same *Long's* "Ancient Custom," (a slave woman painting the eyebrows of a handsome Egyptian girl) very pretty. " The Music Lesson," by *Leighton*—a lovely young woman teaching a lovely little girl to play upon some instrument—very fascinating.

Friday, June 8th.

A very fine warm day.

We dined with Lord and Lady Hanmer; met
Kinglake, Lord Crewe, two Miss Kenyons, Lord
and Lady Hammond.

————

Saturday, June 9th.

A most beautiful day.

We went—invited by Mr. Roundell—to an afternoon party at Dulwich College. It was extremely
pleasant—very unlike the generality of afternoon
parties in London. The company met in the
gallery itself, in the midst of a collection of pictures,
a collection particularly well arranged, very select,
and to be seen most comfortably; and from thence
we strolled out whenever we were inclined, into
a very pretty garden. The day was beautiful, and
everything enjoyable.

In the garden, some very fine Judas trees, the
largest I remember to have seen in England, in full
blossom; also an uncommonly large Catalpa. It is
strange that I had never before seen the Dulwich
Gallery. I recognised several pictures which I had
known by engravings:—*Murillo's* delightful "Beggar
Boys," — his "Virgin and Child;" *Velazquez's*
"Philip IV.; the St. Sebastian (*Guido's*) celebrated
in Alton Locke; the Venus, also mentioned in
Alton Locke, and which I have heard Kingsley
praise, seems to have been removed, though it is
in the catalogue.

I was much struck with *Rembrandt's* "Jacob's
Dream"—the finest picture of his, I think, that

I have ever seen; noticed also some very fine 1877.
Cuyps; some excellent Dutch Landscapes by
Teniers; an admirable woodland scene by *Berchem*,
(Nicholas Berchem).

Gainsborough's portrait of the two Miss Linleys
(Mrs. Sheridan and Mrs. Tickell), very attractive
and interesting.

We met many friends and acquaintances: Aberdare, Sir John and Lady Kennaway, Sir George
and Lady Young, Mr. Eddis, Lord and Lady
Tollemache, Mrs. Grey and Miss Shirreff.

—— ——

Monday, June 11th.

Extremely hot.

Mr. Lofts came in the morning, and we settled
with him for care of my London property.

Leopold and Lady Mary Powys, Mdlle. Van
Randwyck, the Morewoods, and Mrs. Richmond,
came to luncheon—afterwards we went with Mdlle.
V. R. to the National Gallery.

——

Wednesday, June 13th.

We went into the Royal Academy, but had not
much time to spare for it. Admired *Millais'* "Sound
of Many Waters," and *V. Cole's* " Arundel," as
much as the first time. In the former the truth and
beauty of detail in the rocks and vegetation of the
foreground, together with the autumnal colouring of
the woods, are admirable. The foreground will
bear geological and botanical examination; the

1877. lamination of the rocks, their various tints, the patches of lichens and deep green mosses on them, the tufts of grass drifted swept by the water, the red fallen leaves are rendered with marvellous truth to nature. In *Cole's* picture, the sunset glow on the clouds, and the reflection in the water, are exquisite.

We dined with the Bishop of Winchester; met the Bishop of Gloucester, Mrs. and Miss Ellicott, the Bishop (suffragan) of Guildford, Sir John and Lady Duckworth and others. I sat by Lady Duckworth, a very agreeable and accomplished person, a niece or cousin of Sir Charles Lemon; and we had much pleasant talk about Cornwall, as well as other matters. My neighbour on the other side was Mrs Harold Browne, a very pleasant person, particularly genial and kindly.

Thursday, June 14th.

Our dinner party :—the Charles Hoares, Mr. and Mrs. Maskelyne, Rose Kingsley, Mr. and Mrs. Coltman, Agnes Kinloch, Mr. William Nicholson and his daughter, Mr. Medlycott, Captain Medlycott, Minnie, Edward, Frank Lyell, Clement, Harry, Miss Head. I sat between Kate Hoare and Mrs. Maskelyne—both charming.

Friday, June 15th.

Drove with Fanny—we looked into the Abbey. Dear Rose Kingsley dined quietly alone with us—a great delight to me. She is as admirable as ever; there are few women who interest me more, or of whom I have a higher opinion. She is going 1877. abroad almost immediately with a cousin and friend —going to Murren in the Bernese Oberland ; I trust it will do her a great deal of good, and I am sure she will enjoy it. She has never seen the Alps.

We visited Mr. Eddis, and he showed us his studio, and, in particular, several of his charming pictures of children, in which he particularly excels. Also an excellent portrait of Lord Coleridge. Mr. Eddis is a remarkably pleasant man.

June 16th.

I observe in *The Times* the death of Mrs. Norton —for one knows her much better by that name than of Lady Stirling Maxwell; death following very close to her second marriage. Poor woman, in spite of her extraordinary gifts of beauty and genius, she seems to have had far from a happy life, judging from her writings, and one cannot help feeling sorry that this second marriage, which really seemed to promise happiness, should have been so short-lived.

We went to call on Lady Lucy Grant, and to see Miss Grant's studio. She devotes herself in earnest to statuary, for which she certainly shows a remarkable talent. We saw in her studio (besides busts) a very noble figure of St. Margaret ; one of Lady Macbeth, also fine,—and a very lovely group of two children, one asleep, the other holding a violin, and leaning against a pile of musical instruments. Miss Grant is a very pleasing person.

1877.   Afterwards we went for a while to the Zoological Gardens, where, in this splendid weather, most of the animals, especially the birds, were in a state of great animation, and apparently of enjoyment.— Jaguar—a remarkable specimen, very richly coloured and considerably darker than usual, a sort of intermediate grade between the Black Jaguar and the common variety. Bronze-winged Pigeon, from Australia; colouring of its wing-plumage singularly beautiful; a female bird, now setting on its nest, in a bush in the western Aviary.

———

Sunday, June 17th.

Beautiful weather — very hot.   We went to luncheon with the Charles Hoares, to meet there dear Sarah Sanford, newly come to London; and we spent about two hours most delightfully with them.   Kate's dear little boy, a fine little fellow of about one and a half years old, also made his appearance and amused us much.   Sarah is looking much better than when we were with her last year at Woody Bay, and in much better spirits, much more like her former self.

———

Monday, June 18th.

Splendid weather—very hot.

Mr. Lushington, Mr. Robeson, Willie Bruce, Arthur Lyell, came to luncheon.

A visit from Mrs. Berners.   We dined with the Youngs.

Again to the Royal Academy, and saw it to my satisfaction. My impression of *Millais's* and *Cole's* pictures, which I have already mentioned, was confirmed. Admired a picture by *E. Crofts* (a new name to me)—"Ironsides returning after sacking a Cavalier's house;"—very well and carefully painted, telling its story well, with a great deal of expression and character in the figures, though they are on a small scale. The same artist has another picture —"Cromwell at Marston Moor"—well imagined, but not, I think so good as this.

"The Hay Field;" *W. Linnell*—a beautiful South English landscape, with a soft, distant back-ground ; very much to my taste.

"Cherry Blossoms," by *T. Linnell*—a very pretty bit of woodland scenery.

"Friends in rough weather."—*Hook;* a dog bringing ashore through a heavy surf the rope from a boat in danger—capital.

Tuesday, June 19th.

Fine weather continuing.

We went with Minnie to Lord Hertford's, to hear Mr. Brandram read or rather recite, "The Tempest." I was much pleased. He must have a remarkable power of memory, for I saw no book or paper, certainly he did not consult any, and he never seemed for a moment at a loss. His recitation was extremely good, always perfectly clear, spirited, and various, not overdone.

Our dinner:—Mr. and Lady Muriel Boyle, Lord

1877. and Lady Hanmer, the Louis Mallets, Mr. and Mrs. Bonham Carter, Colonel Ducane, Mary Lyell, Minnie, Edward, William Napier, Lady Head.

I sat by Lady Muriel, and was charmed with her.

A great many more guests—"a drum" in the evening.

Wednesday, June 20th.

William Napier came to luncheon. We went (Minnie with us) to Kew, to the Hooker's garden party. The weather being beautiful (as it has been almost continually since the 7th), was very well suited to such a party, for the gardens were in full beauty, with something still of vernal freshness ; only the rhododendrons are passing away.

Sarah and Mrs. Sanford, Nellie Sanford, and Caroline (Truey) Hervey, dined with us, a quiet party of seven (with ourselves and Harry). A delightful evening. Sarah quite charming—quite as she was in her best days, before her ill health.

Thursday, June 21st.

An interesting letter from Mr. Mark Napier.

The Bishop of Bath and Wells, the Morewoods, Minnie Powys, Sir John Kennaway, came to luncheon.

We dined with the Locke Kings.

Friday, June 22nd.

Visited Lady Lilford.

Our dinner party : — the Sanfords (including

Nellie), Colonel and Mrs. Legge, Minnie Powys, 1877.
Minnie Napier, William Napier, Mrs. Forbes,
Mr. Cyril Graham, Mr. Bentham, the Leonard
Lyells, Charles Hoare, William Hoare, Edward.

I sat between Sarah Sanford and Mrs. Legge:
the former I need not characterize; the latter
remarkably pretty.

<hr/>

Saturday, June 23rd.

Fine, but a cold wind.

Kate came to luncheon.

Saw a strong muster of Volunteers in front of the
Athenæum.

Dear little Charlie Seymour was brought by
his nurse to spend part of the day with us. He is a
beautiful and engaging child.

<hr/>

Sunday, June 24th.

Very fine, but cold.

Read prayers with Fanny and Minnie.

Visited Mrs. Douglas Galton; very agreeable
as usual.

<hr/>

June 25th.

Our dinner party (our last in London this season):
—the Bishop of Ely, the Albert Seymours, Leopold
and Lady Mary Powys, the Sanfords (2), the Mac-
Murdos (3), Minnie, Lord Waveney, the Nugents (3),
the Angersteins (2), Colonel Ives, Edward, and
Harry;—22, including ourselves. A few more in the
evening.

Tuesday, June 26th.

A farewell visit from dear Sarah.

Sally and Clement dined with us.

———

Wednesday, June 27th.

A very fine and very hot day.

Farewell to dear Kate Hoare and Minnie.

Down to Melford Hall, Sir Wm. Parker's, by 4.20 train from Liverpool Street. Lord Waveney and Mr. Martyn in same carriage with us, and also staying at Sir Wm. Parker's.

———

Thursday, June 28th.

At Melford Hall.

An excessively hot day.

We went with Lady Parker into Sudbury to the Agricultural Show, others of the family going earlier and others later; suffocating heat and very little to amuse us.

———

Friday, June 29th.

Another extremely hot day. We remained quietly and comfortably at Melford, in pleasant idleness, with the Parkers, who are very pleasant.

Sir William showed me the Church.

Lord Waveney and Mr. Martyn had gone away in the morning.

———

Saturday, June 30th.

We returned home by road, in our own carriage, a pleasant drive of just two hours. All well, thank God.

July 7th 1877.

Barton.

This place is in great beauty. The trees in full rich foliage, and, the season being a very late one, we have now a great many of the flowers which are usually past before the end of June. When we first came home, just a week ago, the lawns looked very parched up; but on the 1st, there was heavy and continued rain, and it has rained more or less, most days this week, to the great rejoicing of the farmers. It is unlucky indeed for the hay making, which has been cut short in the middle; but the long continued drought in May and June—dry cold in May and dry heat in June—has been such an impediment and disadvantage to all farming operations, that the rain is welcomed as a boon. The gardeners are equally pleased with it. All wall fruits, and even apples, are said to have been almost entirely destroyed by the frost in the early part of May, but strawberries are abundant and fine. Roses in great beauty.

LETTER.

Barton, July, 7th, '77.

My Dearest Cissy,

We are now quietly settled at home, for (I hope) a month or more, and I can try to answer your kind letter of the 23rd June. Fanny, who is a superlative correspondent, has already answered the business part of it, and told you what we have

1877. thought it best to settle with regard to Harry, and which I hope you will approve of.

We spent about two months in London, and saw a good many people whom I was very glad to see, besides a great many mere acquaintances; had many very pleasant luncheon and dinner parties at our own house, and some pleasant parties elsewhere; but we saw nothing else besides human beings. In fact a good bit of the early part of the season in London was in a measure wasted, owing to the detestable weather and to Fanny's ill-health (she was so much depressed by an obstinate influenza cold from which, however, a few days at St. Leonards quite relieved her).

I do not think that London suits me so well as when I was younger, for I can no longer take long street walks, and I do not enjoy the fashionable occupation of dowagering in a carriage to leave cards. However, after all, I did very much enjoy some of the society we had in London—only my dear friends the Sanfords (Sarah Hervey and her husband) did not come till a few days before we left town. Well, we left London on the 27th June, and went down to Melford Hall, Sir William Parker's, to which we had been engaged to go on the occasion of a county agricultural show.

The show was a bore, as such things usually are —but our stay at Melford Hall was pleasant enough. Sir William is a great invalid, a victim to gout and leads a very retired, secluded life, so that though our homes are only about fourteen miles apart, I was hardly acquainted with him. But he

is a very agreeable, accomplished, well-read man, 1877.
cheerful, in spite of his infirmities, has seen many
countries, and draws extremely well : so that he is
a real acquisition to our acquaintance. Lady
Parker is very pleasant. They have a family of
only eleven children. Their house is old and
picturesque, built about the time of Queen Mary
(Tudor).

We came home on the 30th, and Leonora and
her two daughters came to us on the 4th of this
month, and are with us now, and will stay, I hope,
till the end of July. The girls are uncommonly nice
girls. I do not know whether Fanny has told you
that we saw *Mrs. Lambert* in London. I was
surprised one day by receiving a note in pencil,
almost illegible, from her, which moved my pity
by mentioning that she was completely blind. We
went to the hotel and had a long talk—that is, she
talked and I answered her questions as well as I
could, and Fanny listened. She talked with
immense vehemence. She did not stay long in
London, and we did not see her again ; and though
I felt pity for her being poor, deaf and blind, I
cannot say that I much wished to cultivate her
acquaintance further.

(*July* 18*th*). We have now a numerous family
party and a very pleasant one : Leonora and her
two girls, Katharine and Rosamond, Emily (Mimi)
and Louisa MacMurdo, and Harry. The girls are a
charming bevy, and all very merry together. Mimi,
one of the most charming girls I know. Arthur

1877. Lyell came to us from Cambridge for three days, and went back yesterday. The other Arthur (*our* Arthur) will come, I believe, to-morrow.

Last week was warm and beautiful, but the last few days have been so wet, stormy, dark and chilly as to be almost wintry. I am afraid you have fallen in with similar change of weather since you left Bellagio, and that it must make you anxious about dear Emily; I trust her health will not suffer, and that you will be able to keep her well and happy through all changes of place and of weather, till you are again established at Mentone.

To day is Fanny's 63rd birthday, and the girls presented her this morning with a beautifully-made wreath of wild flowers, and an exceedingly pretty address, signed by the five. We propose to remain here till the 14th August, then to go to London for a couple of days, and thence to visit the Morewoods (Patience Hervey and her husband) in Derbyshire.

Ever your affectionate brother,

(With much love to dear Emily),

CHARLES J. F. BUNBURY.

JOURNAL.

Monday, July 16th.

Lady Gage, with her party from Hengrave (Mr. Boehm the sculptor, Miss Rhoda Broughton the novelist, and several others), came to luncheon and to see our pictures.

Thursday, July 19th.

Fanny and most of the party went to luncheon at Hengrave.

Dear Arthur arrived.

Fanny received a letter from Mary Harrison.

———

Monday, July 23rd.

Mr. Moutgomerie and his cousin, and a Mr. Smith, came to luncheon and spent the afternoon here.

———

Tuesday, July 24th.

Mr. and Mrs. Walrond arrived.

Dear Katharine and Rosamond left us for Cromer.

For nearly three weeks we have had a very pleasant family party ;—Leonora and her two girls all the time ; Katharine and Rosamond since the 13th ; Emily (Mimi) and Louisa MacMurdo since the 14th ;—a charming bevy of girls. Mrs. Laurie and her niece, Arthur Lyell and Edward, for shorter periods. (By the way, Mrs. Laurie was quite a new acquaintance to me at least ; she is a friend of Isabel Hervey, and a very pleasant person).

The Russian advance seems to be at last becoming rapid and formidable. They have actually passed the Balkan in considerable force, and by a pass which was not supposed to be practicable ; and they have gained an additional passage over the Danube ; and all this time they have fought no considerable action. As far as we can yet judge,

1877. the Russian commanders seem to have shown remarkable boldness and activity, the Turkish, remarkable inertness.

———

Wednesday, July 25th.

Walked through the garden, arboretum, &c., with Mr. and Mrs. Walrond.

Mr. Bentham arrived, and I again walked round the grounds with him.

Lady Susan Milbank, Lady Gage, Mr. and Mrs. Hawkins, Mr. and Miss Bevan and others dined with us.

The joyful news, by a telegram from Charles Hoare, of dear Kate, that she and the baby are doing well.

———

Thursday, July 26th.

Very fine.

We drove to Hengrave with Leonora, Mr. Bentham and the Walronds, and saw the house. Mr. and Mrs. Robeson arrived, also Annie and Finetta Campbell and two of their brothers.

———

Friday, July 27th.

A beautiful day.

A long talk on Mildenhall business with Mr. Robeson and Fanny.

Mr. Bevan and two of his daughters, and Mrs. Glyn, came to luncheon and spent the afternoon with us, and were all very pleasant.

———

Saturday, July 28th. 1877.

A visit from Archdeacon Chapman, and a long talk with him on Mildenhall business.

———

July 29th.

Mr. Bentham is now 77 years old and does not *look* younger, but he seems in good health, sees and hears well, walks well, has an excellent memory; his mental faculties appear to be in as good condition as can be, and his zeal for botany and all things connected with it as keen as if he were young. He is a very agreeable man; kindly and genial, with a great range and variety of knowledge, not at all limited to botany or even to biology; has travelled in almost every country of Europe, and known a great many eminent men. Mr. Bentham admired many of our trees, especially the Catalpa in the arboretum, the Magnolia acuminata, some of the Plane trees, the Cephalonian Fir in the arboretum, the Pinus excelsa, the Menzies Fir.

———

Tuesday, July 31st.

A very hot day.

The assizes at Bury; an unusually heavy calendar, and several extraordinary crimes. I was as usual, Foreman of the Grand Jury; a long and very fatiguing day's work.

Dined with the Judges, Sir Balliol Brett and Sir James Stephen. Both look like thoroughly strong men, both in body and mind.

This was one of the hottest days, if not the very hottest, that we have yet had this year.

Wednesday, August 1st.

The two Judges and their Marshals dined with
us—also Lady Hoste, Mr. and Miss Bevan, the
George Blakes, Mr. Bulwer, Mr. Tyrell.

Dear Leonora and her two girls went away—uncommonly nice girls they are; thoroughly well
educated, intelligent, pleasing in manners, lively and
lady-like.

———

Thursday, August 2nd.

A letter from dear Mrs. Kingsley to Fanny, telling
the good news which has evidently filled her with
joy, about her son-in-law, William Harrison.   Lord
Spencer has given him the living of Brington in
Northamptonshire—the parish in which Althorp is
situated, a good one in point of emolument, and
most agreeable in every way, in short it seems to be
everything that can be desired, and I am most
heartily glad for our dear friend's sake.

———

Saturday, August 4th.

Mrs. Robeson and Mr. Livingstone came to
luncheon, and spent the afternoon here—important
conversation with Mr. Livingstone about Mildenhall.

———

Sunday, August 5th.

We went to morning Church and received the

Sacrament. Mr. Smith officiated and preached, 1877.
being newly returned from Switzerland.

Monday, August 6th.

The meeting of the Bury Temperance Societies
in our park and grounds—a very large assemblage
and great merriment.

Tuesday, August 7th.

Fanny and the young Macs went to Mildenhall
and returned to dinner.

A letter from Mr. Livingstone, accepting the
living of Mildenhall, so I hope that business is at
last settled. That piece of patronage has given me
a great deal of trouble. It was about the beginning
of May, immediately after our arrival in London,
that Mr. Robeson informed me that he meant to
resign the Vicarage of Mildenhall, having been
offered the living of Tewkesbury, by the Chancellor.
During the three years that he has lived at
Mildenhall, he has been a most valuable clergyman,
and done an immensity of good ; indeed he has
quite placed Mildenhall on a new footing ; so that
it is a real misfortune to lose him. And Mildenhall
is a difficult parish to provide for.

Mr. Livingstone, who has now accepted it, has
been Mr. Robeson's successor at Forthampton. He
paid us a visit with Mrs. Robeson on the 4th, and
spent some hours here ; I was favourably impressed
by his appearance, manners, conversation, and
the sentiments he expressed. I have, I think I may

1877. say, honestly laboured to find a good man for this important place, and I wish I may have been successful.

———

Furious storms of rain.

Mr. Courtenay Boyle and Lady Muriel arrived, also Agnes Wilson and Mr. Murray.

———

Distressing news of Sarah Sanford—letter to me from Mr. Sanford, which I answered. Lord John Hervey and Mr. and Mrs. Robeson arrived.

———

Long talk on Mildenhall business with Mr. Robeson and Scott.

———

Mr. Courtenay Boyle went away early,—Lady Muriel after breakfast.

Lord Arthur Hervey came to luncheon. A most agreeable surprise.

———

Very fine and warm.

Up to Eaton Place—we two with the two Mac-Murdo girls, Harry and Arthur. John Herbert came to see us—the MacMurdo girls went away.

We have been in great anxiety about our dear 1877. Sarah Sanford, who has had a serious attack of illness. She was with her husband at Woody Bay; her father and mother were coming into Suffolk, and we were hoping to receive them here on the 9th, when that very morning, we had a letter from Caroline Hervey, telling us that they had been summoned to Woody Bay by a telegram from Mr. Sanford. In the afternoon of the same day (9th), I had a letter from Mr. Sanford himself, giving a fuller account of Sarah. She had for a good while been suffering from a troublesome cough, which alarmed her husband very much. The doctor however gave him comfort, finding that the lungs were all sound. Every day's account since, has been favourable; and she has been going then on so well that her father thought it safe for him to leave Woody Bay and come to Ickworth.

We received the sad news of the death of Sarah Seymour's little baby-girl, Violet.

———

August 15th.

We saw the MacMurdos, Minnie Boileau, and John Herbert. I visited dear Kate Hoare.

———

August 16th.

From London to Alfreton Hall (the Palmer Morewoods) in Derbyshire, by Midland Railway; leaving St. Pancras station at 11.30; waiting an hour at Trent station, and arrived at Alfreton

1877. station a little before 5.   We passed St. Albans,
Luton, Bedford, Kettering, Market Harborough,
Leicester, Loughborough, crossed the Trent near
Kegworth, changed trains at Trent station, a little
way north of the river.

The country not in general interesting till between Leicester and Loughborough, where the bold,
syenitic hills of the Mount Sorrel (or Charnwood
Forest) district, rising pretty abruptly out of the
plain, are well seen on the left.   A little beyond
Trent station we enter on the coal formation, and
continue on it to Alfreton:—a pretty country,
green, well wooded, and of varied surface, though
not rising into very conspicuous hills; often disfigured by collieries and black smoke.

The succession of zones of country of different
geological formations is not so apparent on this line
as on some others; the chalk especially being much
masked by drift.

August 17th—24th.

We stayed at Alfreton with the Palmer Morewoods, spending our time very pleasantly, in spite
of the very bad weather.   Patience is a charming
hostess, and both she and her husband were
indefatigable in their kindness and attention to us.
It added very much to our enjoyment, that Lord
and Lady Arthur Hervey, and Caroline (or Truey)
with them, were staying there at the same time,
and we enjoyed a great deal of their society.

There were other pleasant guests also,—Captain

and Mrs. Elliot, Mr. and Mrs. Hamilton Anson, 1877.
Mr. Chandos Pole. Mrs. Elliot is an aunt of
Charles Morewood—a very good-natured, pleasant
person. Captain Elliot a very agreeable man.

I had wished very much to re-visit Alfreton,
having stayed there with my father and mother, in
1820, when I was just 11 years old, and when the
great kindness shown me by the old Mr. and Mrs.
Morewood of that time made a lasting impression
on me. It was very amusing and interesting to call
up the shadowy recollections of those early days,
and try to compare my ideas with the reality. As
usually happens in such cases, I found the reality
less magnificent than the indistinct images in my
memory.

The old Mr. Morewood who was so kind to me in
1820, had been a great friend of my grandfather,
and there is in the house a very large collection of
my grandfather's drawings and of prints from his
designs, which Patience showed us one evening.
She has also in her own sitting-room, some remarkably fine specimens of his drawings, which she
found in the house, and which are now framed and
hung up.

Some good pictures at Alfreton, especially full
length portraits of Mr. and Mrs. Morewood, by
Romney—excellent. A Salvator Rosa, a very
characteristic specimen of his style; the subject
appears meant to be a sacred one, but it is some
time before one finds out that it represents anything
but some peasants and fishermen with a boat, in a
rocky cove. A landscape with Cows, by *Gains*

1877. *borough*--very good. A portrait said to be of the
Duchess of Kingston, as a flower-girl, I should think
the name very doubtful.

In the morning walked through the gardens with
the ladies.—Fine trees ; a superb sycamore near
the house, I think the finest I have ever seen. A
"fern-leaved" beech (Fagus sylvatica, var. asplen-
ifolia), remarkably large and fine, a perfect dome of
verdure. Rhododendrons grow most luxuriantly
here, forming extensive thickets, must be glorious
when in flower, A very pretty formal flower-garden,
sloping to south, and framed on three sides by
woods or high shrubberies. Kitchen garden also
pretty and old-fashioned. A good deal of glass,
but no large houses. Pinery. The first pine
apples I ever saw or tasted were at Alfreton, in
1820.

In the afternoon Charles Morewood drove us and
others of the party to Wingfield Castle, returning
another way by a very pretty circuit through the
country lying west of Alfreton. Wingfield Castle
(one of the places in which Mary Queen of Scots is
said to have been imprisoned), is a very picturesque
ruin, in a noble situation standing on the brow of a
commanding height, having on one side a steep
grassy descent into a deep valley, while from other
points of view its towers are seen rising out of tufted
trees which cluster on the slope. The Castle is
said to have been built in Henry the Sixth's time,

and to be of the transition style between thorough castle and thorough mansion. Its battlemented walls and high towers have the true castellated character, and on the other hand there are large and beautiful mullioned windows which look as if they belonged to the Tudor time.

The country very pretty, hilly, green, varied and smiling, hills steep but not broken, a great prevalence of pasture, beautifully green, abundance of wood and fine trees. The broad-leaved elm, Ulmus montana, remarkably abundant, and of most vigorous growth.

––––

August 18th.

Very bad weather.

We went with the Morewoods (a large party in two carriages), to Newstead Abbey, which is about 10 miles from Alfreton, in a direction slightly to the south of east. The country, the first part of the way, rather in the same style as that described on the 17th, but not so pretty. Crossed some steep ranges of hill. On this road we pass by Clay Cross, now so much noted for the coal it supplies to the railway.

Near Kirkby, enter on a considerable width of table-land, bleak, bare and barren-looking, though enclosed. The table-land seems to be connected with the "Hills of Annesley." Thence a long descent into a wooded valley, in which stands Newstead. Just outside the gate of the park stands a grand old oak, indeed immense, called the

1877. Pilgrim's Oak, said to be the only one spared, of all the trees about the place, by "the wicked Lord."

Newstead Abbey lies low, as described in "Don Juan," at one end of the lake, in a small valley between wooded hills of moderate slope. The park—what we saw of it—is pretty, with much fern and heath, and woods mostly planted by Colonel Wildman. The house is extremely interesting, especially the relics of Byron, and the beautiful east end of the Church (adjoining the front of the inhabited house), with the great window (now quite empty), and the Virgin and Child in a niche over the window, celebrated in some beautiful stanzas of "Don Juan."

From the entrance door of the house we descend into a very low hall, with a vaulted roof supported on short thick pillars,—it seems a sort of crypt, or undercroft, reminding us of that in the Palace at Wells. Byron's bed-room, with its furniture, prints, &c. (far from splendid or elegant) preserved exactly as in his time, is one of the most interesting parts of the house. Other interesting relics are,—his writing table, at which he wrote the first cantos of "Childe Harold," — a piece cut out of the bark of a beech tree, on which he had carved (in 1813) his own name and that of his sister; the small sword with which the 5th Lord Byron killed Mr. Chaworth in a duel.

There is the small square court, surrounded by cloisters (looking rather dark and damp when we saw it) ; and in the centre the quaint old fountain,

celebrated in " Don Juan," and well represented in 1877.
the illustrated edition of Byron's Works.

The old refectory (?) now a pleasant saloon, of
great height and size, and gorgeously furnished. A
beautiful little chapel (? belonging to the old part of
the house) in a style reminding me of some parts of
Salisbury.

The house was almost re-built (as I understand)
by Colonel Wildman. From such a rapid visit, I
cannot at all understand the plan of it, but it seems
to be very large and handsome. There are very
long galleries, which are now ornamented (as various
parts of the house are) with great quantities of
stuffed exotic birds in fine preservation, and many
other specimens of natural history, especially from
Africa. These, I suppose, were all collected by Mr.
Webb.

Heavy rain prevented us from going into the
garden. The lake is drained for the present, for the
purpose of cleaning it out—which very much injures
the view.

Mr. Webb has been a mighty hunter in Africa,
was intimate with Livingstone, and accompanied
him in some of his earlier expeditions. A room in
the house is called Livingstone's room, and there
are some other relics of the great missionary
traveller.

We returned to Alfreton in heavy rain, partly by
a different road ; passed some extensive iron-works,
and saw the flames rising from the furnaces.

Alfreton.

Morning extremely wet.   In afternoon, walked
with most of the party to Swanwick Colliery, a mile
or two from Alfreton, in the valley below.   No work
going on, of course, but we saw the mouth of the
pit, the engine house, the great piles of coal, sorted
according to its quality, and the huge mounds of
shale and rubbish.   The pit belongs to Charles
Morewood.   He showed me the various qualities of
the coal.   Most of it is rather anthracitic in appearance, being compact, dull and stony-looking; but
there are veins and layers of a crystalline, bright,
shining, bituminous coal; and as I understand him,
the best quality is that in which these two varieties
alternate in somewhat regular layers.

We had a good view of Alfreton Hall from near a
wind-mill on a hill on the south side of the valley.
It stands very well, on the ridge of the hill, screened
but not concealed or smothered by rich woods,
between which the bright green pastures slope down
into the valley.—The town adjoining to it,—side by
side with the park,—but well screened off by the
woods; very conspicuously red in the general view,
nearly all the houses being of red brick.

————

We went with the Morewoods and Caroline
Hervey (Charles Morewood driving), to Hardwick
Hall, about seven miles from Alfreton, to the north-
east or thereabouts on the borders of Nottingham-

shire. A long and steep ascent through the park
to the brow of the height on which the house stands.
Oak trees very numerous, a great many of them
apparently very old, with their upper branches dead.

Ruins of the old Hall, at a little distance from
the house, on rather a lower level. Supposed to be
of Henry the Seventh's time (Murray).

Part of the park of Hardwick is in Nottingham-
shire.

Hardwick is a very fine stately and interesting
old Elizabethan house, remarkably well preserved.
Situation fine :— standing on a high ridge of hill,
rising above everything near it, and seen against the
sky, it looks uncommonly majestic. Front very
peculiar and striking in appearance, from the extraordinary number and size of the windows ; it is
really "more window than wall ;" and the windows
(at least those of the three lower stories) are suc-
cessively larger in each story upwards. A very high,
curious and elaborate open-work parapet all along
the top, with the letters E. S. (Elizabeth Shrews-
bury, the notorious Bess of Hardwick, the jailer of
Mary Stuart), of gigantic size, often repeated.
Interior of the house, remarkable for the complete-
ness and exactness with which the old furniture and
arrangements of the sixteenth century are preserved
in every particular. It seems to show us exactly
how our ancestors lived—at least the *great people.*
Gray in his letters characterizes Hardwick very
well :—

"One would think Mary Queen of Scots was
"but just walked down into the park with her guard

1877. "half-an-hour. Her gallery, her room of audience, "her ante-chamber with the very canopies, chair "of state, footstool, *lit-de-repos*, oratory, carpets, "and hangings, just as she left them. A little "tattered indeed, but the more venerable, and all "preserved with religious care."—*(Gray to Dr. Wharton, from Cambridge, December 4th, 1762).*

The profusion of well-preserved tapestry in all the rooms is quite extraordinary: I never before saw anything like the quantity. The hall very stately, panelled up to a certain height with tapestry above, a wide gallery along one side, the walls adorned with deer's antlers and armour. At one end of the hall, a beautiful statue of Queen Mary Stuart, by *Westmacott:* the inscription under it is copied from that which Johnson composed at the request of Boswell.* The great gallery magnificent: 169ft. long (extending along the whole east front), abundantly lighted, and its walls absolutely covered with portraits—many curious. In the council chamber, and I think also in the library, the lower part of the walls covered with tapestry, the upper part with stucco, bearing coloured figures in relief, very curious.

There is an extensive and fine view from the roof. The garden—a fine example of the old formal style, with clipt evergreen hedges.

---

* See Boswell's Johnson.—letter from Johnson of July 4, 1774.
" *Maria Scotorum Regina, nata* 1542, *a suis in exsilium acta* 1567, *ab hospitâ necidata* 1587."

A very agreeable excursion to Matlock, with Charles Morewood, the Bishop and Caroline, Mr. and Mrs. Hamilton Anson, going by one road and returning by another. Each way we cross high and very steep hills, which, as we approach Matlock, become more broken and intersected by picturesque little valleys.

Matlock and Matlock Bath have both now grown into real towns; both stand on the margin of the Derwent: the first on the east bank, and higher up than the other, which is on the opposite side. We crossed the river by a bridge, and went down the right bank to one of the hotels at Matlock Bath, where we left the carriage for a time. Beautiful woods along the river side, especially on the east bank, and superb limestone cliffs towering above them. The High Tor especially fine. I botanized for awhile in the woods on the east bank, but with indifferent success: it was too late in the year.

The weather indeed was favourable for mosses, but it was not the proper season for the fructification of most of them.

The way back to Alfreton especially beautiful; at first along the margin of the Derwent, southwards, to Cromford Bridge; past Mr. Arkwright's beautiful park and grounds, an amphitheatre of brilliant verdure enclosed by rich woods,—then up a very long and steep hill and over the bleak and open heights through the village of Crich, and down a long descent into the valley below Alfreton, passing

1877. by the fine ruins of Wingfield Castle. The Derwent at Matlock, is very rapid, but very turbid, and it is not much less so at Chatsworth.

———

LETTER.

My dear Katharine,

Many thanks for your letter of the 17th, and for the papers enclosed in it. I will take great care of these papers, and return them to you as soon as I have copied the list of mosses, which will be very useful to me.

Shall you want the papers before we return to London, about the middle of September? In M. Schimper's list of Brazillian mosses, 1 recognize almost all the names as belonging to species which I also found in that country.

We arrived here last Thursday, the 16th (after a very good journey from town) and have been spending our time most pleasantly. Mrs. Morewood is a charming hostess, and both she and her husband are most kind, and do everything to make our time pass agreeably. Lord and Lady Arthur, and their unmarried daughter are here, and add much to the agreeableness of the party; and there have been, and are also some other pleasant people. The unfavourable weather indeed has been something of a drawback to our enjoyment, for it has rained more or less every day since we came and sometimes very heavily. But in spite of this, through the kindness of the Morewoods, we have seen Wingfield Castle, Newstead Abbey (particularly interesting to me), Hardwick Hall, very curious, and full of historical interest, and Matlock—beautiful. The whole drive between this and Matlock is full of beauty.

The last accounts of dear Sarah Sanford are good, though her recovery is slow ; but I am afraid she will require much care for a long time, though it is a comfort that the lungs are not attacked. I am afraid my niece Susie (Cecil's wife) is again danger-ously ill. though we have only heard indirectly as yet of her illness, and know no particulars. I am grieved for poor dear Sarah and Albert losing their baby,—it is a great sorrow to them and to Minnie. I am very sorry also to hear of the death of George Loch—he was a valuable man, and much to be regretted, though it is not to me as it is to Fanny the, loss of an old friend.

We ourselves, I am thankful to say are at present quite well.

Much love to Rosamond,

Believe me ever,

Your affectionate brother,

CHARLES J. F. BUNBURY.

JOURNAL.

1877. We went with the Morewoods and Arthur
Herveys to a garden party at Mr. Strutt's, at
Belper: pretty gardens, in a fine situation.
Conifers thriving particularly.

We met here Mrs. Douglas Galton and Gwendoline.

———

We said farewell, with much regret, to our kind
friends at Alfreton, and (sending our servants and
luggage by railway), went in an open carriage to the
Chatsworth Hotel at Edensor, at the very gate of
Chatsworth Park.

From Alfreton to Edensor, 17 miles.

First over the high hills to Matlock, the same
way we went on the 21st. Thence a delightful
drive up the beautiful valley of the Derwent. Hills
on both sides steep and high, beautifully variegated
with wood and pasture, and here and there abrupt
faces of limestone rock.

General characters of the country along the
Derwent:—great prevalence of grass land, pastures
beautifully green, fields rather small than large,
abundance of hedge-row timber, and of scattered
trees, often large. The hills, though very steep,
generally enclosed almost to their tops, but here
and there one sees tracts of moorland.

Darley — very picturesque wooded knolls and rocks.

A beautiful drive through part of Chatsworth Park, before reaching Edensor; good view of the house, backed by a magnificent bank of wood, over which rise the bleak tops of the hills. Trees grow very finely in this limestone district; the most characteristic and remarkable, both for abundance and beauty, are the Ash and the large-leaved Elm (Ulmus montana).

August 25th.

Weather very cold and wet, in spite of which we managed to see Bakewell Church and Haddon Hall.

Bakewell, beautifully situated, in a fertile green valley, through which flows the pretty little River Wye (a tributary of the Derwent) between high and very bold hills, richly wooded, but with bare and moory tops. The Church stands well, on high ground, over-looking the town; it is modern for the most part (rebuilt in 1841), but the west end of the old Norman Church is preserved, with a remarkable Norman doorway. Many curious tombs and coffin lids carved in very peculiar style, and looking very ancient, were discovered when the church was re-built, and are now placed in the porch. In the interior several curious monuments of the Foljambe, Vernon, and Manners families.

(Bakewell Church). Very large marble tomb of Sir George Vernon and his two wives; another of

1877. his daughters and heiress Dorothy, who eloped with Sir John Manners, and carried the Derbyshire estates of her family to the Mannerses; also a large marble tomb of Sir John Manners, the son of Dorothy. Mural monument to Sir Godfrey Foljambe and his wife, curious as showing the style of armour and dress of the period (the latter part of the fourteenth century).

Haddon Hall. A fine stately old house, somewhat in the style of Hardwick, but deserted and unfurnished, though not ruinous. The situation very picturesque; amidst rich woods, on the brow of a height which rises abruptly from the flat, green meadows, through which the pretty little Wye flows in an intricately winding course.

The house stands on steep ground, so that the outer and inner courts, and the basement storeys of the two sides, are on different levels. Gardens full of mournful yew trees and clipped evergreens, terraced and balustraded also on different levels. One part of these gardens known as Dorothy Vernon's Walk; and the small door opening on it from the ball-room, as Dorothy Vernon's door, by which she escaped to her lover, Sir John Manners. The ball-room or great gallery, very long—the walls panelled — great bow window adorned with heraldic devices. The great hall or banqueting room, with its gallery for musicians, and the stags' antlers decorating the walls; the close and immediate communication with the kitchen also remarkable. The smaller dining room or withdrawing room, low, with coved ceiling, and

panelled walls, on which are portraits of Henry VII.
and his Queen, and Will Somers the jester. The
steps leading to this room are formed each of a solid
log. Much tapestry in many of the rooms; some
fine, the greater part much faded.

On the whole, somehow, a melancholy impression
is produced on us by the interior of Haddon Hall,
in its present state. Though one cannot exactly
say that it is neglected, there is an indefinable look
of death and decay about it, much more than was
observed at Hardwick.

Sunday, August 26th.

We went to morning service in the pretty Church
of Edensor, just within the gate of the park.

In the Church of Edensor, is a very large and
showy monument to the first Earl of Devonshire
(d. 1625).

In the afternoon we drove for a good while about
Chatsworth Park—very fine. Trees magnificent,
especially the broad-leaved elm (Ulmus montana),
ash, lime, and sweet-chesnut; the two latter, no
doubt, planted, the others, indigenous.

We saw a heron flying over the river.

The park seems to occupy the whole valley of the
Derwent for some distance, extending to and over
the crests of the hills on each side.

August 27th.

Weather abominable, but we saw the interior of
the house of Chatsworth — magnificent; a true

1877. palace. I cannot pretend to describe it, or to mention more than a very few of the fine things it contains.*

A most admirable collection of original drawings by the great masters; this alone would occupy one many days, or weeks, in its study.

Wonderful carvings in wood, by Grinling Gibbons and others—almost inconceivably delicate. Ceilings painted by Verorio, Laguerre, and Sir James Thornhill, in the style immortalized by Pope.

The profusion and beauty of the various kinds of marble and alabaster in all the rooms, are astonishing; many from Derbyshire, others Italian or antique.

In one of the rooms a special collection of minerals, apparently very choice, obtained and arranged (we were told) by the late Duke. Among them a remarkably fine crystal of emerald.

Pictures :—

The famous portrait of the beautiful Duchess of Devonshire (Georgiana) with her child on her lap—*Reynolds*—lovely. This is in the private apartments.

A sketch of Lady Spencer, and her daughter the beautiful Duchess, as a child—*Reynolds*.

A Family Group, very fine—by *Titian* (or according to Waagen, by *Paris Bordone*).

A Rabbi, by *Rembrandt*.

Several fine *Vandycks*.

In the Sculpture Gallery, which is remarkably

---

* There is a very good description ot Chatsworth, in Waagen's " Art and Artists in England." (1838), volume 3., 224.

beautiful ;—A very fine colossal head of Napoleon's 1877.
Mother, by *Canova* (imitated from the antique
Agrippina) ; *Canova's* Endymion ; *Gibson's* Mars
and Cupid ; *Thorwaldsen's* Venus.

In another room, *Canova's* lovely Hebe.

August 28th.

We went over the hills to Buxton*, in an open
carriage, an interesting drive. At first, followed the
same road as to Bakewell, turning off a little short
of that town. Pass through Ashford, a neat little
town, in a very pretty situation ; here many marble
shops, the fine Derbyshire marbles being quarried
hard by. One of the kinds of marble found near
Ashford, is a very fine uniform black ; another, a
rich, deep red, very like the *rosso antico* ; others
variegated. A grey kind, variegated in a curious
and interesting manner by the sections of the
Encrini, of which it is almost made up, is used for
chimney pieces in the hotel at Edensor.

Thence, ascend the course of the little river Wye,
through a picturesque pass or *dale*, narrow and
winding ; the hills rising very steeply from the
margins of the rapid and impetuous stream ; their
sides clothed either with grass or brushwood ; in
many places, abrupt bare walls and towers of grey
limestone rock, crowning the heights or standing
out through the brushwood.

After ascending a long way through this beautiful
pass, we quit the stream, and come into a more
open country, high and bleak, where hedges are

* In the visits we paid this summer, and during the whole of the journey,
we were accompanied by Arthur MacMurdo.

1877. entirely superseded by walls of loose stones. Many
of these walls have an odd appearance, from the
mixture of pieces of blackish *toadstone* (amygdaloid)
with the pale grey limestone. In this high
country, also, as far as I could see, all the hills
appeared green and grassy; I saw no heath. Near
the highest grounds we pass through the village of
Taddington, grey and solid looking. A long
descent again into the valley of the Wye, for the
road cuts off one of the great windings of the river
by striking across the hills on which Taddington
stands. We saw the amygdaloid *in situ* in one
place as we returned down the hill from Taddington; it had the peculiar nodular look often observed
in trap rocks, especially when beginning to decom-
pose. Thence to Buxton up the course of the same
river, through another beautiful winding dale,
perhaps less wooded and more rocky than that by
which we had ascended on the other side. Fine
sections of the carboniferous limestone beds, some
natural, some made in the operations for the
railway.

From Edensor to Buxton, by this road, 15 miles.

Geranium pratense in great abundance and
beauty between Ashford and Taddington, by the
side of the Wye, on the grassy margins of the road,
and at the foot of the stone walls.

The Wye rises among the hills behind Buxton.

Though we saw a few Ferns (of which Cystopteris
fragilis was the most remarkable), I saw nothing
like the abundance of them with which we were so
much struck, last year, in Devonshire.

A violent storm came on just as we entered 1877. Buxton, but it had cleared off before we started on our return.

The only plant *savouring* of a mountainous locality which I observed, besides the Cystopteris, was Saxifraga hypnoides.

The limestone rocks in some parts were richly clothed with mosses, but seemingly in little variety; the most abundant, by far, appeared to be Hypnum molluscum and Leskea sericea.

————

We saw the gardens at Chatsworth—they well deserve their fame. The great palm-house is finer in the general interior view than that of Kew; the clear, uninterrupted, far extending avenue of lofty tropical trees, visible in all its length, is glorious. I have nowhere else, in Europe, seen tropical trees looking so much at *home*—so nearly as they do in their own country. The palms and bananas are magnificent; the leaves of the latter of wonderful size and beauty. But, vast as is the height of the house, some of the palms have grown till they touch the roof, and one has actually lost its head through being too tall. The tree ferns I did not think very remarkable. Smaller ferns in the utmost luxuriance and profusion, as if in their native places, clothing large dead trunks and branches of trees with a most beautiful drapery.

The great palm-house is 65ft. high, 126ft. wide, and 276ft. long.

1877. Musa coccinea, in flower, most beautiful.

Araucaria Braziliana, the largest I have seen in Europe, nearly as high as the house, but with scarcely room enough for its enormous spread of branches.

In some of the smaller hothouses, we saw numerous species of Sarracenia—more numerous and varied than I ever saw before—all thriving, and very curious and beautiful.

In the kitchen gardens, in another part of the park, at some distance, are two small hothouses, one appropriated to the Victoria and other tropical aquatics, the other to the Amherstia. In this last house the heat excessive. The Amherstia, we were told, had flowered well this last spring. It is quite a tree, with branches gracefully drooping all round it, and fine, large, healthy, handsome foliage. In the same house with this, were Nepenthes of several species, in great beauty and luxuriance.

Out-of-doors, we admired the really gigantic and very picturesque rock-work near the great palm-house, and the beautiful views of the park from various points and in various directions: also many individual trees, especially a chesnut (Castanea vesca), perhaps the finest I have seen in England, measuring 15ft. round the trunk at 3ft. from the ground, and high in proportion.

————

August 30th.

We drove from Edensor to Rowsley Station, thence by railway to Derby: drove from the station

there to Derwent Bank, where the Galtons received 1877.
us very pleasantly. It is a pretty spot, though
overlooking a town (Derby), which is by no means
lovely.

Galton showed me the Derby Arboretum, and
also the Exhibition of Fine Arts which is now open.
The Aboretum, given to the town by Mr. Joseph
Strutt, and laid out by Loudon, is a pretty public
garden, but not now kept up with any scientific
aim.

At the Exhibition we saw many good pictures
lent by gentlemen in the country: in particular,
several excellent portraits of Wright of Derby,* an
artist with whose works I was before unacquainted,
but whose merits seem very great. One portrait by
him, of a Lady Wilmot, I thought particularly
clever. The Duke of Devonshire had sent some
fine pictures from Chatsworth, and portraits from
Hardwick.

———

Saturday, September 1st.

From Derby, by railway, to Bettisfield station,
a few miles beyond Whitchurch, where Lord
Hanmer's carriage met us. Passed Tutbury station :
—just beyond it, the ruins of Tutbury Castle are
seen rising picturesquely above the trees, on the top
of a bold wooded hill.

---

* Wright, Joseph, distinguished by the name of Wright of Derby, at which
place he was born in 1734. He was a scholar of Thomas Hudson, at the
same time with Mortimer. He was elected an Associate of the Royal
Academy in 1782, but resigned his diploma in disgust, at a Mr. Garvey being
elected R.A. before him. He died in 1797.—*Bryan's Dictionary of Painters.*

1877. From Derby, nearly to Longton, the country new red sandstone, nearly a plain, green, rich, and well wooded. A sudden transition, a little before reaching Longton, from the new red to the coal. New red at one end of a tunnel, coal at the other. Here begins the potteries district, not extensive, but intensely marked with all the features of a colliery district :—coal and iron pits, huge engines, clouds of smoke, huge piles of rubbish. We passed Stoke-upon-Trent—notorious for Dr. Kenealy.

From Crewe to Bettisfield, the new red again, a green and smiling country.

Monday, September 3rd.

At Bettisfield.

Incessant and heavy rain till near evening, when I had a short walk with Lord Hanmer.

Mr. Henry Hanmer and Mr. and Miss Kenyon, came to stay at the house.

Tuesday, September 4th.

Lord Hanmer took us to see the church of Hanmer, about two miles from Bettisfield, in a very pretty situation, at one end of the *Mere*, which is a very pretty sheet of water, surrounded by rich bright green slopes, partly wooded. The church, a solid, venerable, and rather handsome one, with a battlemented square tower, stands at the top of the slope at one end of the mere. Within are monuments to many of the Hanmer family ; particularly a very large one to Sir Thomas Hanmer, the Speaker, and editor of Shakspeare, with a very 1877. long and pompous Latin epitaph, by Dr. Friend, who was celebrated by Pope for the prolixity of his epitaphs.—

> "Friend, for thy epitaphs I'm griev'd,
> Where still so much is said,
> One half will never be believ'd,
> The other never read."

Also a large monument to the first Lord Kenyon, with another long inscription, but in English.

September 5th.

We had a long drive with Lord Hanmer, Mr. Kenyon, and his niece Miss Kenyon, through a pretty country, into the valley of the Dee, where it forms the boundary of Flint and Denbigh shires; crossing the river into Denbigh by Bangor Bridge, and recrossing into Flint by Overton Bridge, a few miles off. Bangor, a mere village, was the seat in the middle ages of a great and famous monastery. Nothing now remains of this, but its site was in the low alluvial meadows beside the river, on the Denbighshire side.

This part of the country (on both sides of the Dee) is not an absolute plain like the neighbourhood of Chester, still less is it like the limestone hill country of Derbyshire; it is in fact very uneven, so that hardly two fields together are on the same level; and in driving about it, one is continually going up or down; and the slopes are pretty steep too. But all the hill-tops are pretty much on the

1877. same general level — none are conspicuous; the valleys are narrow, the slopes smooth and unbroken.

On the Flintshire side the ground is high, and there is a considerable and steep descent to the bridge. Here we saw a man on the river in a *coracle*, a very ancient sort of boat, made of wicker, covered with leather or some waterproof material, and looking very like a tub. When he landed he took the coracle on his back, and then looked like a tortoise in its shell.

The Dee is here a fine river, with a considerable body of water, rapid and turbulent, of a dark brown colour.

The name of the Dee, as Lord Hanmer told us, is derived from its colour, *Dee* (which is the same as *Dhu*), signifying dark or black.

There is a beautiful view at Overton Bridge, where the banks are high and steep, and richly wooded, with superb trees ; oak, sycamore, and lime, hanging over the rushing stream. The town of Overton, on the brow above the bridge, is one of the Flintshire boroughs.

Not far from hence, on the high ground, is a cemetery, from which there is an extensive and very beautiful view ; a rich, green, smiling valley, almost in the form of an amphitheatre, with the Dee winding through it in graceful curves ; and in the background, seen over the hills in which this valley is embosomed, the mountains of North Wales, the mountains above Llangollen, I believe, and still further off, a part of the Berwyns. They are visible

also from Lord Hanmer's grounds ; and in another 1877.
direction is seen the Wrekin.

These mountains, indeed, we had in view during
nearly the whole of our drive.

Thursday, September 6th.

We went with Lord Hanmer and Sir Robert and
Lady Cunliffe, to see Oteley, Mr. Mainwaring's, on
Ellesmere Lake—a fine new house, in a delightful
situation ; on the brow of a high bank, overlooking
the whole extent of Ellesmere, a beautiful sheet of
water, much larger than the mere at Hanmer, with
richly wooded banks ; the Church-tower of Ellesmere, a very fine and conspicuous object in the
view, directly opposite to Oteley, at the other end
of the lake.

Mr. Mainwaring's gardens, beautiful ; in terraces
descending one below another down the steep slope
to the lake, the terraces laid out in flower beds, and
flanked by fine flourishing trees. Conifers and
evergreens, very fine. A beautiful conservatory,
full of splendid tropical plants—communicating with
the house.

We visited also a new Church, at Welsh Hampton, built by the same Mr. Mainwaring — very
elaborate and richly decorated.

Swans on Ellesmere—their beautiful flight—the
first time I remember to have seen them on the wing.

Sir Robert and Lady Cunliffe were guests at
Bettisfield during the latter part of our stay there,
both of them very intelligent and agreeable ; Sir
Robert particularly so.

Friday, September 7th.

We drove with Lord and Lady Hanmer, Lady
Cunliffe and Mr. Henry Hanmer: saw some farms
and cottages of Lord Hanmer's: returned by
Hanmer, where Fanny made some purchases.

————

September 8th.

We left Bettisfield; went by railway to Chester
by Whitchurch Malpas, and Tattenhall.

In the afternoon we drove to Eaton, and saw the
gardens, but they fell far below our expectations.

————

September 9th, Sunday.

Chester.

In the morning we attended service in the
Cathedral: in afternoon we walked on the top of
the walls all round the town.

Chester is still a remarkably curious and
picturesque old city—the most remarkable, I think,
that I have seen in England. The old style has
been well kept up where repairs and restorations
have been necessary, and the curious arrangement
of the *Rows*—I was glad to see—still exists to a
considerable extent.*

The town has spread widely beyond the walls in
extensive suburbs, so that a great part of what we
see from the walls is no longer country but
covered with houses. Observed a great change in
this respect, since '46, when we spent a day here in
our way back from Dolgelley.

* Chester. The Rows are well described in the *English Cyclopedia*.

Tower—included in the wall—on which Charles 1877 the First stood and saw the defeat of his army on Rowton Heath — a fact commemorated by an inscription.

* * *

September 10th.

We spent the morning in examining the Cathedral and St. John's Church, and the afternoon in an expedition to Beeston Castle, 10 miles distant. The Cathedral is fine, but I should not place it in the first rank of those I have seen. There is a very picturesque view of it from the top of the city wall, on the east. The exterior having been much decayed and dilapidated owing to the action of the weather on the red sandstone of which it is built, extensive restorations have been carried out at a very great expense under the direction of the present Dean, but they are not yet completed.

Beeston Castle is very picturesque and remarkable ; a true hill-fort, standing on a singularly bold and isolated hill. This hill is evidently an outlier of the Peckforton range, from the extremity of which it is separated by only a short space ; but it is singularly abrupt and detached, the ground immediately around it on all sides being completely a plain. On one side (the side on which Arthur and I walked up) the ascent is easy, only moderately steep, covered with fine, short grass, like the Malvern hills ; on the other sides, all the upper part of the hill is quite precipitous, standing up in abrupt, rugged cliffs,

1877. of bare sandstone rock, very bold and picturesque. Below the cliffs, the hill-side is very steep, but accessible, and covered with trees, in great part evidently planted. The rock is red sandstone, similar in grain, colour, stratification, and general appearance, to the ordinary sandstone of the country, but must be much harder.

The remains of the Castle on the summit and on the accessible side of the hill, are not very considerable, yet picturesque.*

The Peckforton hills, of which this is an outlier, are on a long ridge with a wavy outline, rising boldly out of the very level and fertile plain of the New Red, and form an agreeable and interesting feature in the landscape. Peckforton Castle (modern).

Lord Tollemache's great house stands high on the extremity of the range facing Beeston, above thick woods, and looks very well.

Tuesday, September 11th.

From Chester to London by the North Western : had a good journey, and arrived safe and well—God be thanked.

The MacMurdos (including Emily and Louisa), dined with us.

We were occupied in town till the 17th September, when we went to visit William Napier, at the Government House, Sandhurst. We passed that evening very pleasantly, but unluckily we were tied

---

* The height of Beeston Castle hill is 366ft. above the sea.—*English Cyclopaedia.*

down by an engagement in London, so that we could afford but a little time to Sandhurst. The 18th, we drove over (William lending us his carriage and horses), to Farnham Castle, to visit the Bishop of Winchester and Mrs. Browne, and spent the afternoon with them. They were very kind and friendly to us, and very agreeable.

The drive from Sandhurst was very pleasant, across the fine open heath country by Farnboro' and Aldershott, and through the beautiful region of hop cultivation about Farnham, with the high hills of the Lower Greensand in the back-ground. The hop harvest was just beginning, and the plants in their utmost beauty.

Farnham Castle, crowning the summit of an abrupt hill, rising high above the town which clusters around its base, and overlooking the rich landscape far and wide, has a stately and noble appearance. The remains of the keep (part of the still existing masonry of which, I think, is believed to be of Stephen's or Henry II.'s time), as well as those of the outside walls, and the great ditch exterior to these, are very interesting and impressive. The gardens, which were famous in the time of the late Bishop, as well as of Bishop North, (See the Linnean Transactions, v. 6., p. 312), are no longer kept up in the same splendour (the See being so much poorer), but are beautiful from their situation ; terrace above terrace of gay flowers, and luxuriant climbers mantling the grey old walls.

That same day (September 18th), William gave a dinner party, at which we met several pleasant

1877. people: in particular, Sir Garnet Wolseley, whose conversation I found very agreeable and interesting. The 19th we returned to London, and the 22nd we came home. All well, thank God.

LETTER.

48, Eaton Place, S.W.,
September 14th. 77.

My Dear Katharine,

I return herewith the list of Indian mosses, with many thanks, it will be a great help to me, for I have made a copy of it.

Fanny's journals have given you such complete and accurate accounts of all our movements and sight-seeings, that I need not add anything. I had few opportunities of botanizing, except in the drive from Edensor to Buxton over the limestone hills, and then I found nothing new to me, though I enjoyed the looking out for plants and the seeing again the beautiful Geranium pratense and Cystopteris fragilis in their native localities. I am sorry I have no specimens of the Pimpinella magna for you; I thoughtlessly and stupidly neglected it, and the one or two specimens I did collect, were quite withered when we got back to Edensor.

I was delighted with Chatsworth, both house and gardens, especially the latter: and there is also a collection of minerals in the house, which I should have liked mightily to look over at my leisure.

We are now quiet here for a little while, devoted

to the interesting occupation of shopping; we are going to William Napier's on Monday for a day or two, and hope to return to Barton on the 22nd, to-morrow week.

We had a very pleasant dinner with the Mac-Murdos yesterday, and met Kate Ambrose, whom I was very glad to see, not having met her for several years.

Ever your affectionate brother,

CHARLES J. F. BUNBURY.

JOURNAL.

Wednesday, September 19th.

The Louis Mallets, the MacMurdos (3), and Sir Francis Doyle dined with us—all very pleasant.

Saturday, September 22nd.

Down to Barton.

Sunday, September 23rd.

Read 7 chapters of Job : also Kingsley's sermon on Christ weeping over Jerusalem.

Our anxiety about dear Sarah Sanford's health is by no means yet removed. Her recovery is very slow; it now appears that one lung is affected, though as yet in a slight degree. The doctors now recommend her spending the winter in a warmer climate, and I believe Madeira is decided upon.

1877.   Monday, September 24th.

The Percy Smiths came to luncheon, and Lady
Hoste called.

———

Wednesday, September, 26th.

A beautiful day.

We dined with Mr. Greene and Lady Hoste:
met Mr. and Mrs. Wilmot Horton, Captain and
Mrs. Horton and Freda, the Thornhills, the George
Blakes and some others—a pleasant party—Clement
went with us.

Read 3 chapters of Job.

———

Friday, September 28th.

My Barton Rent Audit—satisfactory: we both
of us afterwards had luncheon with the tenants.

———

Saturday, September 29th.

Very fine.

Drove into Bury with Fanny: called on Mrs.
Rickards and had a pleasant talk with her—she
looking very well for her age.

From September 22nd till now, November 16th,
we have been quiet at home, with the exception of
a visit of three days to Lord and Lady Gwydyr at
Stoke Park, close to Ipswich. During nearly the
whole time, we have had an almost constant
pleasant succession of company at home :—

Mrs. Ellice and Ellen, Minnie Boileau, Admiral
Spencer, Leopold and Lady Mary Powys, Leonora
and her two girls (only for a very few days) ;

Willoughby Burrell; Lady Head and Amabel; Edward, Montague and Susan and Emily Macmurdo, Katharine and Rosamond Lyell, Mr. and Mrs. Montgomerie, Captain and Lady Edith Adeane, Lady Gwydyr and Miss Burrell, Sir John Shelley, Captain Crosbie; these last fourteen or fifteen came expressly for the balls at Bury, and for our dance, on the 23rd, 25th and 26th of October.

Minnie Powys, General and Mrs. Lynedock Gardiner, Edward Campbell with his daughters, Annie and Finetta, William Napier, the Edward Goodlakes, Mrs. Wilson and Agnes, and lastly the Matthew Arnolds, Lady Rayleigh, Richard Strutt, Mr. and Mrs. John Paley.

Now all our guests are scattered, and since the 13th we have been alone with dear Minnie Napier, who is not like a guest, but one of ourselves—truly a sister.

———

Saturday, November 17th.

Fanny had a very interesting letter from Mr. Sanford, from Madeira—on the whole a tolerable account of dear Sarah.

Lady Hoste, Agnes Wilson, and the Prideaux Brunes came to luncheon.

———

Monday, November 19th.

Very cold; thermometer 26 degrees. Mr. Greene, Mr. King and Mr. Denton came, and we discussed the question of the roads.

———

Tuesday, November 20th.

Cold rain. Our *migration* into first-fioor rooms was completed.

————

Wednesday, November 21st.

Very cold and damp, ending in rain.

Fanny had a letter from Lady Arthur, with continued good accounts of dear Sarah.

Lord Waveney, Lady Hoste, Mr. and Miss Prideaux Brune came to luncheon, and stayed long.

————

Friday, November 23rd.

Our accounts of dear Sarah Sanford have been very interesting, and though by no means such as to free us from all anxiety, their general tendency is comforting. When we returned home (September 22nd) she was still at Woody Bay, and there was great cause for anxiety as to whether she would be able to bear the removal to Dartmouth, where it was intended that she should embark for Madeira. Happily that removal was effected without apparent harm ; but the accounts we received from Dartmouth were hardly cheering. However, on the 4th of this month came the welcome news (in a letter from John Hervey to Fanny) that Sarah had been safely carried on board without appearing to suffer, and had started in fine weather, and with favourable prospects. And on the 8th we were made very happy by the information, also from John Hervey, that the family at Wells had received a telegram

many blessings which have been bestowed on me. It is especially a cause of thankfulness that my dear wife and myself have enjoyed good health, and are as firmly knit together in love as ever. And though one of the most precious of our friends has been removed to a better world, we still have the good fortune to possess many who are worthy of all our love and esteem.

Cissy and her daughter Emily are spending this winter as they did the last, at Mentone; and although Emmie has had no serious attack of illness, yet I fear, from the accounts we have, that her health is not thoroughly re-established. On the other hand, Susie (Cecil's wife) who has for years been in a precarious state of health, and has repeatedly been in extreme danger from hæmorrhage of the lungs, seems lately to have rallied in a surprising manner, and to be now in comparative good health. Last June, anyone would have said that she was much nearer to death than Sarah Sanford. Beside Sarah I do not know that we have lost any intimate friends this year. George Lock should perhaps be excepted; we both were very sorry for his death. Our good servant, Edgar Calthorpe should also be mentioned. Two are gone whom I knew very well in former times: Sir Henry Codrington, with whom I was familiarly acquainted when we both were young; Lady Northampton (Eliza Elliot) whom I saw very often while I was at the Cape, her father being the Admiral on the station, and her sister and herself the particular friends of Cissy and her sister.

1877.   I do not think I had seen Lady Northampton since my return from the Cape, but it was always agreeable to me to remember her, as she was associated in my mind with the memory of the happy year I passed there.

I could not number Mr. Motley among our friends, though I have seen him formerly at the Lyells'; but I think he was the greatest *public* loss in all this year's obituary.

Another remarkable man with whom I was once, long ago slightly acquainted, was Mr. Henry Fox Talbot.

Some remarkable women have died this year:— Mrs. Norton (Lady Stirling Maxwell) whom I remember seeing long ago in the splendour of her beauty. Lady Smith (the widow of the great botanist, Sir James), who died in her 104th year; Mrs. Chisholm and Miss Mary Carpenter. The other celebrated names in the obituary of this year are Thiers, Brigham Young, Jung Bahadoor (the Viceroy of Nepaul), Le Verrier, General Changarnier, and General Aurelle de Paladines.

Canon Beadon of Wells, whom we saw there in 1870 and 1871, is still alive, and completed his *hundredth* year on the — day of this month of December. It is said that the Queen wrote to congratulate him on the occasion.

Of my reading in the course of the past year, I can give but a poor account; I have been desultory and irresolute, beginning various books which I have not gone through. Those that I am best satisfied with, are—"The Life of the Prince Consort," the 2nd and 3rd volumes (the last, indeed, I have not yet 1877. finished), and several of Burke's works.

So ends 1877.

God grant that the coming year may be better spent by me.

Deo Gratias.

BOOKS READ.

1877.

Life of the Prince Consort. Vols. 2 & 3 (the last not yet quite finished.

Burke. On the Cause of Present Discontents ;—on American Taxation ;—on Economical Reform.

Lecky. History of Rationalism (second time). Arnold's Lectures on Modern History (third or fourth time).

Autobiography of Harriet Martineau.

Froude. Short Studies on Great Subjects, third series.

Life of Sir Ralph Abercrombie. (Very poor indeed).

Lives of Simon de Montfort and of the Black Prince.

Life of Lord Shelbourne, volume 2.

A. R. Wallace's Address to the Biological Section of the British Association at Glasgow (excellent).

Alfred Newton's Address to the Sub-section of Botany and Zoology, at some meeting (striking and interesting).

1877. Bentham on Classification of Monocotyledons, in *Linnean Society's Journal*, v. 15.

The Disowned, by Lord Lytton (Bulwer)—a powerful and interesting novel, but disfigured by very poor attempts at humour.

"Mr. Smith," by Mrs. Walford.   A clever and very amusing novel.